DEMOCRATIZATION OF EXPERTISE?
Exploring Novel Forms of Scientific Advice in Political Decision-Making

Sociology of the Sciences

VOLUME XXIV

Managing Editor:

Peter Weingart, *Universität Bielefeld, Germany*

Editorial Board:

Yaron Ezrahi, *The Israel Democracy Institute, Jerusalem, Israel*
Ulrike Felt, *Institute für Wissenschaftstheorie und Wissenschaftsforschung, Vienna, Austria*
Michael Hagner, *Max-Planck-Institut für Wissenschaftsgeschichte, Berlin, Germany*
Stephen H. Hilgartner, *Cornell University, Ithaca, U.S.A.*
Sheila Jasanoff, *Harvard University, Cambridge, MA, U.S.A.*
Sabine Maasen, *Wissenschaftsforschung/Wissenschaftssoziologie, Basel, Switzerland*
Everett Mendelsohn, *Harvard University, Cambridge, MA, U.S.A.*
Helga Nowotny, *ETH Zürich, Zürich, Switzerland*
Hans-Joerg Rheinberger, *Max-Planck Institut für Wissenschaftsgeschichte, Berlin, Germany*
Terry Shinn, *GEMAS Maison des Sciences de l'Homme, Paris, France*
Richard D. Whitley, *Manchester Business School, University of Manchester, United Kingdom*
Björn Wittrock, *SCASSS, Uppsala, Sweden*

For further volumes:
http://www.springer.com/series/6566

DEMOCRATIZATION OF EXPERTISE?

Exploring Novel Forms of Scientific Advice in Political Decision-Making

Edited by

SABINE MAASEN
University of Basel, Switzerland

and

PETER WEINGART
University of Bielefeld, Germany

Library of Congress Control Number: 2009932133

ISBN 978-1-4020-4698-8 (PB)
ISBN 978-1-4020-3753-5 (HB)
ISBN 978-1-4020-3754-2 (e-book)

Published by Springer,
P.O. Box 17, 3300 AA Dordrecht, The Netherlands.

www.springer.com

Printed on acid-free paper

All Rights Reserved
© Springer Science+Business Media B.V. 2009
No part of this work may be reproduced, stored in a retrieval system, or transmitted
in any form or by any means, electronic, mechanical, photocopying, microfilming, recording
or otherwise, without written permission from the Publisher, with the exception
of any material supplied specifically for the purpose of being entered
and executed on a computer system, for exclusive use by the purchaser of the work.

TABLE OF CONTENTS

PREFACE vii

Chapter 1: WHAT'S NEW IN SCIENTIFIC ADVICE TO POLITICS?
Introductory Essay 1
Sabine Maasen and Peter Weingart

Chapter 2: BIOETHICAL CONTROVERSIES AND POLICY ADVICE: THE PRODUCTION
OF ETHICAL EXPERTISE AND ITS ROLE IN THE SUBSTANTIATION OF
POLITICAL DECISION-MAKING 21
Alexander Bogner and Wolfgang Menz

Chapter 3: ADVISORY SYSTEMS IN PLURALISTIC KNOWLEDGE SOCIETIES:
A CRITERIA-BASED TYPOLOGY TO ASSESS AND OPTIMIZE
ENVIRONMENTAL POLICY ADVICE 41
Harald Heinrichs

Chapter 4: INSTITUTIONAL DESIGN FOR SOCIALLY ROBUST KNOWLEDGE: THE
NATIONAL TOXICOLOGY PROGRAM'S REPORT ON CARCINOGENS 63
David H. Guston

Chapter 5: REPRESENTATION, EXPERTISE, AND THE GERMAN PARLIAMENT:
A COMPARISON OF THREE ADVISORY INSTITUTIONS 81
Mark B. Brown, Justus Lentsch and Peter Weingart

Chapter 6: EXPERTISE AND POLITICAL RESPONSIBILITY:
THE *COLUMBIA* SHUTTLE CATASTROPHE 101
Stephen Turner

Chapter 7: KNOWLEDGE AND DECISION-MAKING 123
Frank Nullmeier

Chapter 8: SCIENCE/POLICY BOUNDARIES: A CHANGING DIVISION OF LABOUR
IN DUTCH EXPERT POLICY ADVICE 135
Willem Halffman and Rob Hoppe

Chapter 9: INSERTING THE PUBLIC INTO SCIENCE 153
Heather Douglas

TABLE OF CONTENTS

Chapter 10: BETWEEN POLICY AND POLITICS
Or: Whatever Do Weapons of Mass Destruction Have to Do With GM Crops? The UK's GM Nation Public Debate as an Example of Participatory Governance 171
Simon Joss

Chapter 11: PARTICIPATION AS KNOWLEDGE PRODUCTION AND THE LIMITS OF DEMOCRACY 189
Matthijs Hisschemöller

Chapter 12: JUDGMENT UNDER SIEGE: THE THREE-BODY PROBLEM OF EXPERT LEGITIMACY 209
Sheila Jasanoff

LIST OF AUTHORS 225

BIOGRAPHICAL NOTES 227

AUTHOR INDEX 231

PREFACE

Giving sound advice to political decision-makers has become an important task for scientists in modern knowledge-society. However, difficulties seem to be growing. In the election year of 2004, more than 4'000 scientists, including 48 Nobel prize winners, signed a statement that opposed the Bush administration's politicized way of handling scientific advice; we currently see a host of new forums of extended expertise designed to counter and complement purely academic advice; at the same time, various attempts emerge to regulate public deliberations and formally integrate their recommendations into decision-making procedures by elected political bodies at the local, national and supra-national levels. These are but a few indicators of the uneasy relationship between science and politics – the increasing need for reliable knowledge in virtually all policy fields notwithstanding.

Scholars in sociology of science and science policy are empirically inquiring into the novel forms of scientific advice to political decision-making: apart from providing valuable insights into the plurality and intricacy of challenges at the science-politics interface, the studies convened in this volume also ask whether there are ways in-between technocracy and sheer instrumental use of scientific advice in politics. Not surprisingly, there are no easy answers, yet promising approaches, to grasp the problem of how scientific advice and the political demand for transparency and lay participation may be institutionalized so as not to compromise the ethos and standards of science nor the functions and legitimacy of politics.

This book is based on a conference on "Scientific Expertise and Political Decision-Making', held at the Institute for Science Studies at the University of Basel, Switzerland, December 4-6, 2003. We gratefully acknowledge the financial support received from the Schweizer Nationalfonds, Fritz Thyssen Stiftung and the Freiwillige Akademische Gesellschaft, Basel. Moreover, we owe special thanks to Mario Kaiser for his assistance in organzing the conference as well as to Lilo Jegerlehner for diligently handling the manuscripts. Both accompanied the editors throughout the process with utmost competence and care.

January 2005

Sabine Maasen, Basel
Peter Weingart, Bielefeld

CHAPTER 1

SABINE MAASEN AND PETER WEINGART

WHAT'S NEW IN SCIENTIFIC ADVICE TO POLITICS?

Introductory Essay

THE CASE REOPENED

All of a sudden, 'scientific advice to politics,' the 'nature of expertise,' the 'relation between experts and policymakers' emerge as variations of a topic that is the subject of workshops and congresses, of articles and books. Not just the science studies community but political scientists and philosophers of science alike have discovered the theme. It goes without saying that the topic as such is an old one. Once before, namely during the 1960s, there had already been an intense discussion about science and politics. How is this sudden resurrection of the issue to be explained? In what ways is it framed differently? Which of its features remained unchanged?

The literature on the science – politics nexus of the 1960s makes it clear that the overriding concern of the time was the problem of technocracy. It was seen differently in the United States and on the Continent, authors in the US being primarily concerned about the fate of democratic institutions under the growing influence of scientific experts while analysts on the Continent saw the rationalizing impact of science on the often cumbersome democratic mechanisms. In the US, titles like *The Scientific Estate, The New Priesthood*, and *The Scientific Power Elite* were popular reading among scholars of political science and science policy studies (Gilpin and Wright 1965; Lapp 1965; Lakoff 1966; Price 1967). Their writings reflected the fundamentally democratic tradition of political thought distrustful both of the wisdom of experts and the rationality of central political authorities informed by them. On the continent, scholars such as Richta in Chechoslovakia, Schelsky in Germany (Schelsky 1965) as well as a vigorous school in France (e.g., Ellul 1964; Meynaud 1969; Touraine 1971) were much more receptive to the promises of a technocratic rationality in politics. The basic difference between these authors was the conception of the nature of politics vis-à-vis that of science. The American scholars adhered to the Weberian distinction between science and politics that allowed a rationalization of the means of politics while the irrationality of decisions as such was inevitable. In the European context, this was discussed as the dichotomy of technocratic versus decisionist models of scientific advice to politics (Habermas 1962; Marcuse 1961; Weingart 2001: 133). The commonality of these approaches was the conception of science as objective, reliable knowledge.

Sabine Maasen and Peter Weingart (eds.), Democratization of Expertise? Exploring Novel Forms of Scientific Advice in Political Decision-Making – Sociology of the Sciences, vol. 24, 1–19.
© Springer Science+Business Media B.V. 2009

A number of parameters and basic assumptions of this debate have since changed. First of all, some aspects of the political system have undergone a considerable transformation. Most importantly, since the 1960s, the industrialized countries have experienced a further push of *democratization*. This has become visible in the emergence of political movements operating outside the system of formal political institutions whilst their activities exerting an influence on it. The anti-nuclear and environmental movements are the most pertinent examples. They have become the models for a host of other new forms of broadened public participation in various political contexts. 'Round-tables,' moderated discourses and other conflict resolution mechanisms involving policy-makers and citizens have been established as part of the political system albeit outside the constitution.

A second major development has taken place within science: *It has been politicized.* As a result of the public debates over nuclear energy and environmental protection in the 1960s, scientists were drawn into the political process. They were instrumentalized as experts whose technical know-how was to support political positions on both sides in vicious controversies over technical issues. The public appearance of experts defending contradicting positions made it apparent to the public for the first time that scientific knowledge is not unequivocal, that its implementation entails risks, and that there can, in fact, even be a complete lack of knowledge. Experts, it was discovered, are far from representing neutral knowledge but rather interpret the state of research in various, even completely contradictory ways, taking sides with their favored political positions and/or lobbying for their own interests. This resulted in a dramatic loss of authority of scientific experts in their role as political advisors (not so much of science as an institution!) and, more seriously, a change in the perception of scientific knowledge that could no longer be taken as neutral, objective and reliable (Bimber 1996).

A third change refers to the relation between science and politics as well as the society at large: *the democratization of expertise.* The general democratization, the de-mystification of scientific knowledge and of scientists themselves, and the shift towards new public management have resulted in demands addressing the scientific community. The latter is to be held accountable for the public expenditure allocated to it for research. This may be seen as the first phase of the democratization of expertise. In essence, this means that the promise of the eventual societal utility of knowledge production is no longer taken at face value but has come under much closer scrutiny than in the previous 'social contract' between science and society (Guston and Kenniston 1994). In a second phase, scientific expertise has come under the influence of a related demand. The 'democratization of expertise' is the order of the day in national governments and supra-national bodies such as the EU. A general shift is seen to be taking place from a legitimation through knowledge to a legitimation through participation (EU-Commission 2000; Abels 2003). In academic discussions, this development has been accompanied by a discourse on the 'robustness' of knowledge and on the dispersal of sites of knowledge production outside of the established universities and research institutions (Gibbons et al. 1994; Nowotny et al. 2001).

Against the background of these changes, it is safe to assume that the recent debate on the relationship between science and politics cannot be a mere repetition of its predecessors. In fact, while some of the fundamental issues are reoccurring, the debate is now framed differently, with new problems coming into focus. Indeed, basic problems such as the tension between knowledge and power are unlikely to have a 'final' answer. Rather, they are re-phrased to accommodate new conditions and circumstances, which begs the following question: What exactly is new in the arrangement of scientific expertise and political decision-making? While the novelties may indicate new societal demands on knowledge and decision-making, the prime concern of the *new debates* is how reliable knowledge can be made useful for politics and society, at large. More specifically, how can epistemically and ethically sound decisions be achieved without losing democratic legitimacy? After all, technocracy (the dark side of expertise) is still lurking, but nowadays this concern is articulated as the call for 'democratizing expertise.'

Although we find the same implicit fear of the unaccounted rule of experts at the basis of this debate,[1] the perception of scientific knowledge is different. Rather than being perceived as superior, it is seen as *uncertain, risky*, and *incomplete*. With this shift of framework, the new problem is to pay attention to the properties of knowledge itself. What role does it play in the advisory process, how does it assert itself and what impact does it have and how is it affected by the process? More specifically, how can the objective of democratization of expertise be achieved without compromising the quality and reliability of knowledge? Put in different terms, the issue is how to accommodate *social robustness* of knowledge with its *epistemic quality*, how to bring the *legitimacy* of knowledge (and the experts who represent it) in line with its *adequacy* or *epistemic quality*.

Looking at recent literature on these questions, the danger that new simplified polarizations, programmatic romanticisation and wishful thinking prevail over scrupulous analysis seems ubiquitous. In contrast to this stance, we hold that, however the problem is phrased, a fundamental difference between scientific knowledge and political decision-making remains. For example, any attempt to interpret the production of reliable knowledge 'as analogous to politics' fails dismally just as the opposite, namely that politics could be interpreted as the search for reliable knowledge, would seem outright absurd. These thought-experiments seek to point out the fact that the postulates stipulated by the emergence of new modes of knowledge production and the desirability of 'socially robust knowledge' pose very complex epistemological, sociological and institutional questions. This holds for the more practical suggestions of the democratization of expertise as well. While these phenomena are now rightfully attracting the attention of many scholars in science policy and neighboring fields, providing the new focus of the more sophisticated debate, the latter has only just begun.

We contend that the issue is a precise analytical delineation of the differences between science and policymaking, seeking to establish what it means to democratize expertise and to produce reliable knowledge under that assumption. How can policymaking proceed under conditions of uncertain knowledge? What are the institutional solutions, if any, to the dilemma of the democratic illegitimacy of experts? The problem is, thus, twofold, pointing to the consequences of the new constellation for both the epistemic nature of science advice and for the democratic nature of policymaking. It is for that selfsame reason that the problem has attracted the attention of different fields of research, from sociology of science and science policy, to political science and philosophy of science.

It appears most fruitful to conceive of the relationship between science and politics as one between two differentiated subsystems with fundamentally different codes of operation as an analytical frame. While science, as a subsystem, primarily adheres to the code of 'truth,' politics is primarily guided by the code of 'power'. In a nutshell, this claim postulates that, ultimately, science should produce truth, whereas political decisions should safeguard power. Given these basic distinctions, in a knowledge-based society many interactions occur between science and politics. However, this does not lead to an intermingling of codes or subsystems, respectively. Rather, the nature of the relationship between science and politics is one of 'coupling.' For example, if decisions are science-based, they strive to rely upon and legitimate themselves with 'true' knowledge, yet for politics, the truth of the knowledge in question is not a goal in itself but a means to make lasting decisions that keep the decisionmakers in power. While superficial observation seems to suggest a 'blurring' at the interface of science and politics, the analytical specification we favor focuses on the consequences the mutual reference of the systems has for each of them.[2] Expert advice is a case in point for what 'coupling' means in this context.

In the following chapter we will give an overview over the most visible changes at the science-policy interface, to start with, followed by novel attempts at describing those new alignments. Thereafter, we will focus on one particular form of democratizing expertise that has attracted much interest, namely the involvement of public participation in technical decision-making. In our last chapter, we will ask what extended expertise stands for. While most authors focus on challenges regarding legitimacy, we will draw on the issue of increased responsibilization of citizens (O'Malley 1996). Vis-à-vis the ideal of a strong democracy, administering knowledge-based decisions operates by way of involving experts – scientific, political, or lay – in their capacity as responsible citizens. On this note, we will make a case for extended expertise as a governmental technique in neo-liberal societies.

WHICH ARE THE NEW ALIGNMENTS BETWEEN SCIENCE AND POLITICS?

There can be little doubt that in spite of a general consensus on loss of authority of the traditional scientific expert, the reliance of policymakers on expert advice has increased continuously over recent decades. It is probably more correct to speak of a different role of scientific knowledge. It is no longer seen as an unequivocal source of indisputable truth but rather as a necessary resource of policymaking even though it may be contested and open to interpretation in a specific case. If this may be regarded as an ongoing *scientization of politics,* the seeming paradox is, that it parallels a general *democratization of scientific advice* in the specific sense that the expertise of advisers has become accessible to contending groups in the democratic process. This engagement of experts on opposing sides of the political spectrum, focused on specific issues with a wide range of disciplines involved, has revealed to the public whatever differences exist between them and within their respective fields of knowledge. These are amplified by media reporting and, being addressed to the general public, inadvertently result in the *politicization of science.*

While it would seem that under such conditions, knowledge becomes less valuable as a source of legitimacy to support political decisions the evidence is – surprisingly – to the contrary. The number of experts called upon by governments, the numerousness of advisory committees assisting policymakers continue to grow. Within the framework of 'coupling,' however, this is the result to be expected of a mutual dependence of science and politics under contemporary conditions of knowledge production and political decision-making. With the latter increasingly seeking security and legitimation in knowledge, the former is referred to before its certification by the scientific community in ever-more specific contexts (e.g., foresight, evaluation, regulation, ethical deliberation) by 'clients' (e.g., the executive, parliament, NGOs) with ever-more specific interests. Yet, by way of lending specific advice to (mostly less specific) inquiries, scientific knowledge cannot provide the unequivocal answers required by politics. Vis-à-vis these considerations, it seems far less bewildering that the boundaries between purportedly neutral scientific advice and partial decision-making, a vision still upheld by the traditional decisionist model, become much more complex. We would like to elaborate on a few changes only:

One important characteristic of the new alignment between science and politics is the *proliferation of expertise*. This reflects the fact that policy-makers and CEOs in companies alike, strive to legitimate their decisions with reference to knowledge to an increasing degree. Whereas expertise used to be located either in the bureaucracies of governments or in academia, it now covers a broad range of knowledge from various sources. The latter include the social sciences, led by economics which are the most important. Foreign policy is advised by experts on foreign countries, by security advisers, and more recently by 'experts' on terrorism. They may be recruited among political scientists but also among historians, journalists, and scholars of Islam. Foreign aid policies are based on the advice of a plethora of experts on agriculture, land and water management, health and financial systems as well as engineers, relying heavily on the advice of NGOs that have acquired first hand knowledge in dealing with specific issues in developing countries. By the same token, environmental policies receive their backing from climatologists and biologists among others, as well as drawing on experts on forestry and marine life and activists working within local and international groups that have made the protection of the environment their cause. These examples may suffice to point out that, in essence, the expertise sought by decision-makers is not limited to established fields of academic research but reaches beyond its confines into areas of very practical knowledge.

Moreover, *experts abound*. Proliferation of expertise reflects the ongoing specialization of knowledge that leaves virtually everybody to be layman in almost any realm of knowledge and expert in a very narrowly defined area of activity. Hence, practically everybody can be addressed as an expert for something. To notice lay people as 'experts on everyday life' and have them participate in all kinds of collaborative planning and decision-making (e.g., Després et al. 2004) is but the most recent result of this development. What is more, seeking an 'outside opinion' either by scientific, professional or lay experts has become a pattern both in business and in policymaking, and it serves not only to obtain requisite knowledge for the solution of particular problems but also to legitimate decisions that need an authoritative independent reference.

This does not leave the notion of expertise unaffected. The concept of expertise appears to be extended to the point of denoting almost any kind of knowledge. More often than not, it indistinguishable from experience accumulated in the course of pertinent professional activities. Yet, even where it (rightfully) insists on being firmly rooted in scientific knowledge, *expertise is transgressive*: expert scientists have to synthesize all available knowledge and thereby necessarily transgress the boundaries of their discipline, in a first step. Second, they have to address audiences never solely composed of fellow-experts. Their propositions have thus to be sensitive to a wide range of demands and expectations and relate to the heterogeneous experience of mixed audiences (Nowotny 2003: 152).

The expansion of what is taken to be expert knowledge beyond the boundaries of academically established disciplines is also reflected in the range of *institutions and institutional bases of individual experts*. Giving advice to policymakers is not limited to eminent scholars based at universities who translate their reputation in the scientific community into the authority of their counsel. The number and types of institutions involved in the business of producing expert knowledge and from which advice may be sought have broadened dramatically. Think-tanks, be it non-profit or commercial, with varying relationships to science have emerged. Some attempt to establish academic credibility, others distinguish themselves by keeping a distance to academic accreditation claiming to provide knowledge that is of use to policymakers rather than being acceptable by academic standards (Stone and Garnett 1998). More recently, consulting firms whose traditional clientele were business and industry have entered the arena of advice to policymaking. They now counsel governments in the art of new public management.

This pinpoints another new phenomenon: the *explicit recourse to partisan advice* rather than 'objective knowledge.' Yet, the growing acceptance of partial expertise cannot simply be dismissed as a misguided reaction on the part of scientists whose knowledgeable opinion is called for in matters within and outside their expertise. After all, it is the result of the politicization of expertise and its use by different political factions inside and outside governments. The degree to which the different organizations retain their link to the authority of scholarly knowledge varies. Some think-tanks operate as politically neutral knowledge brokers, others are explicitly committed to an ideological program or to political parties. On the one hand, this raises the question if and how the latter can claim any objectivity for the knowledge their advice is based on. On the other hand, their very existence demonstrates that objectivity qua neutrality may not be a major concern anymore, having ceded its place to the exploitation of the realm of interpretability of knowledge. Policymakers are known to favor advice supporting their convictions (Murswieck 1994: 105). Bimber claims a secular tendency towards politicized advice: "... the ascendance of politicized advice, constructed and presented with advocacy in mind" (Bimber 1996: 16). Institutions providing policymakers with knowledge that is ideologically and politically 'reliable' are instrumental in debates with opponents.

At the same time, individual experts and entire advisory panels are called upon quite frequently to *multiply expertise*. Hoping that they neutralize each other, the number of knowledge–based opinions is increased, thus leaving room for the 'genuinely political' decisions (Bogner and Menz, this volume). The downside of this intentional 'flooding' of the political discourse with expert opinions is that the outcome is even harder to control. In other words, the increasing need to legitimate political decisions with knowledge only sharpens the conflict between 'knowledge and power,' resulting in the ever-more strategic instrumentalization of knowledge.

Finally, *expertise has become a commodity*. Advice is no longer given on demand only. A new feature of the political world in all post-industrial or knowledge societies is the multitude of institutions engaged in producing and communicating knowledge, thereby trying to shape the political agenda. The long–established practice of the US National Academy of Sciences to issue public reports on themes of public interest that it chooses without being commissioned by the government has become a widely followed pattern. Knowledge has been turned into a commodity marketed by institutions if only to demonstrate their capacities to respond if called upon for advice in the future or to act strategically in public discourses. One important implication is that knowledge appearing in public, unlike advice given to policymakers in private, cannot be controlled by them anymore. Together with those actors issuing knowledge, the media take a crucial role in diffusing it to the public, selecting, amplifying and thereby shaping it in their own right.

The features of the new alignment between science and politics thus described highlight the dilemma of scientific advice in mass democratic societies. As expert knowledge has grown in importance as a political resource, actors in the political arena attempt to obtain and control the knowledge that is relevant to their objectives. This competition for knowledge, which already represents 'democratization by default,' has resulted in the loss of science's monopoly on pronouncing truths. At the same time, scientific knowledge has often been revealed to be uncertain, ambiguous and incomplete. Intermediate types of knowledge, expertise specifically developed for the solution of particular problems, hence, not generalizable, gain in importance. Accordingly, accomodating the accountability of experts and the reliability (i.e., quality) of knowledge emerges as *the* crucial issue. There are different solutions to that dilemma, some epistemic, others institutional.

INTERACTIONS BETWEEN POLICYMAKERS AND EXPERTS

Unlike other political processes, the interaction between experts and policymakers is shaped by knowledge, among other things, its content and meaning in relation to the interests and objectives of the actors involved. In order to better understand the mechanisms by which expertise becomes transgressive, hence, functional for political decision-making, and based on a systems-theoretical approach (see above), we suggest to clearly differentiate two modes of communication and activity: *knowing* and *deciding*. The mode of science is oriented to the continuation of systematic knowledge production, to learning and, thus, to the questioning of existing knowledge. The mode of politics, by contrast, is oriented to the closure of public conflicts through compromise, using knowledge strategically as it unfolds. Therefore it is valuable if it supports decisions that are the outcome of compromised interests, but it also entails the risk to de-legitimate past or future decisions. Given its inherent qualities, the continued production and appearance of scientific knowledge constantly irritates politics in unpredictable ways. All the more so, the greater the prestige of the knowledge (indicated by the standing of the expert or the institution it originates from) and the more reliable it has been proven to be in solving perceived problems.

In order to advance beyond the crude models of scientific advice to policymakers, recent analyses focus on the real complexities of the advisory process. Numerous empirical studies have generated a multitude of configurations between experts and decision-makers, revealing insights about the motives and interests of those engaged in the process, of the significance of knowledge in determining their behavior, and of the dynamics between them. Above all, they show that the reality of expert advice to policymaking does not lend itself to simple generalizations. It just happens to be too diverse and complex. A first goal may thus be to arrive at descriptive typologies of heuristic value for comparative studies providing initial insights into the design options for organizing advisory systems, given that in pluralistic knowledge societies policy-oriented knowledge communication is not a one-way street any more, having turned into a dialogistic endeavor instead (Heinrichs, this volume).

Obviously the perspective taken also depends to a large extent on the choice of theoretical frameworks. To name but a few: *Systems theory* highlights the differentiation between variegated social systems and the intransigence of their respective operational codes. Thus, systems theoretical accounts focus on communication that reveals the structural 'coupling' of systems. In the case of the science – politics link, this perspective emphasizes the mutual dependencies of advisors and policymakers and the interlocking of politicization and scientization processes which contradict earlier, unilinear models. *Principal – Agent theory* focuses on the translation of political goals into practice. It deals with the problems that emerge in the advisory process, asking how and why knowledge is introduced and how it relates to the political objectives of the principal. More precisely, since accountability is a two-way street, it demands both a responsible agent and a vigilant principal (Jasanoff 2003: 158). Consequently, the following questions arise: first, how is this vigilance to be exercised? Second, 'who will watch the watcher?' (Guston, this volume). *Political theory*, especially *theory of democracy* is concerned with shifting sources of legitimacy. In the case of the advisory process, the issues are the legitimacy of knowledge as a political resource and its strategic use as well as the ways in which interests and values are represented vis-à-vis knowledge. 'Representativeness' and 'resonance' are two criteria that combine a measure of scientific validity with aspects of both participation and leadership elements of democratic representation (Saretzki 1997; Brown et al., this volume).

Needless to say playing out one theoretical approach against another is of little use to. Given the complexity of the phenomenon, they complement each other quite well and, seen side-by-side, provide a broad spectrum of insights into the advisory process. First, all approaches draw attention to the multitude of ways in which the institutions, procedures and discourses involved are arranged in such a way as not to sacrifice one goal for another, i.e., sacrificing quality of knowledge for control of procedures, or representativeness of institutional frameworks for resonance. These concerns are indicative of our basic assumption that *science and politics adhere to different operational logics*. This tension is reflected in various institutional, procedural and communicative arrangements designed, yet never truly managing, to balance the conflicting demands.

Second, analyses increasingly focus on the fate of knowledge in the advisory process, i.e. on the question how *reliable knowledge* enters the policy process, how it is deployed, i.e. used or ignored, and how it is evaluated with reference to the objectives of the actors involved. The example of the Columbia shuttle catastrophe is a case in point, exploring the ambiguities contained in expertise and responsibility. The system adopted by NASA was an attempt to hold those who expressed opinions responsible for their views, as they bore upon their decisions. In the course of the action, it turned out that the attribution of 'whose opinion is correct' and 'who is responsible for finding out' requires meta-expertise (and meta-responsibility, for that matter) which do not exist. Provisional surrogates for meta-expertise, such as discussion and sharing concerns, can lead to ironic consequences: The decisions they engender are products of consensus for which no one is formally responsible (Turner, this volume).

Third, although Don Price already addressed the concern about the advising experts pursuing their own interests without being democratically legitimated, he saw these interests associated with political support of science. The basis of expertise is no longer limited to academic scientists and, thus, the issue of experts' self-interest becomes more diffuse but also more pronounced. Experts are not just pursuing professional interests, as Price feared. They also attempt to have a direct influence on political decisions.

This changes the nature and image of the advisory process: it can be seen as a communication over power (of definition). Policymakers have to control the appearance of knowledge in order to retain their autonomy of decision-making. Accordingly, they will attempt not only to control the use of advice, its publicity or secrecy as well as its interpretation, but also the selection of advisors and the fields of knowledge they represent, for these factors may decide the answers they can expect. This is illustrated by conflicts among different branches of governments, e.g. executive and legislature, over which expertise should be accessible to whom.

Experts, in turn, have a stake in their professional credibility. Advice that goes unheeded may discredit them with their professional communities. They must have an interest in the truth-value of their expertise and, thus, in the attention it receives from the decision-makers. This is elucidated by conflicts among experts over who represents the correct and most pertinent knowledge (Weingart 2003). Corresponding studies based on an integration of *interaction and structural theory* (Nullmeier, this volume) reveal the intricate patterns of interaction between policy-makers and advisors. In this perspective, social structural relations are constituted by acts of ascription that, in turn, are constituted by speech acts and their concomitant validity claims. By way of analyzing micro-discourses at the interface of science and politics, we observe, for example, the emergence of scientists as 'political knowledge entrepreneurs,' who carefully seek acceptance of their knowledge in its empirical, normative and conceptual aspects.

Fourth, many studies draw attention to the fact that the nature and impact of knowledge varies with the kind of decisions it is supposed to inform. In routine regulatory contexts advice normally rests on knowledge that commands a high degree of consensus among experts who more often than not are scientists, as for example in the setting of emission control standards or in the licensing of new pharmaceuticals

(Krücken 1997; Zhou 2002). Because of the routine nature of the decisions and the high degree of consensus over the knowledge communicated in these contexts, these processes are rarely publicised. If, on the other hand, advice is sought on an issue that attracts a great deal of public attention and where the knowledge involved is uncertain and contested among experts the process is much more likely to become politicised. Moreover, media attention adds to this complexity as neither of the parties involved can control it. In recent years, several instances such as the debates over BSE, stem cell research and climate change may serve as illustrations (Weingart et al. 2002). Thus, the more routine the advisory context and the more consensual the knowledge engaged in the process, the less public attention it will receive, and the less politicised it will be.

Last, not least, it ought to be noted that the science-policy boundaries do not change in an instant. Rather, empirical studies emerge who emphasize the discontinuity of those changes because different patterns co-exist, partly competing, partly complementing each other. In the Netherlands, for instance, corporatist, neo-liberal and deliberative arrangements co-exist (Halfman and Hoppe, this volume). Experts represent relevant knowledge either 'behind the scene' or guard the boundaries of the battle-field (corporatist), expertise is coordinated by market structures (neo-liberal) or public decision-making is seen as prime source of collective reasoning and argumentation (deliberative), each arrangement having its specific costs (e.g., lack of transparency, 'contractualization').

In sum: It is precisely the multitude (and combination) of theoretical perspectives that allows to see behind the somewhat 'messy' surface of those new arrangements between science and politics.

NEW FORMS OF REPRESENTING THE PUBLIC AT THE SCIENCE – POLITICS INTERFACE

Perhaps the most challenging research site in looking at the science – politics interface is now the new form in which the public participates in embedding new technologies in society. The call for a 'democratization of expertise' may actually be seen as genealogical descendants from first protests against nuclear energy and the pursuant development of technology assessment (TA) right down to staging 'round tables' discussing the introduction of genetically modified plants among experts and laymen. Recent studies inquire into the intricacies of such participatory instruments (Abels and Bora 2004). While for the most part being in favor of such instruments, they still identify problems and questions that call for further improvements:

Many authors agree that public participation in technical decision-making is able to short-circuit many of the chronic problems in policy-making, including lack of public trust in technical work, lack of empowerment of citizens, and access to reliable data. On the downside of public participation, a number of problems remain, however, particularly concerning the details of such participatory processes (Douglas, this volume). How do participants get involved? How is their role delineated? Who sets the agenda? Yet another type of problem arises with respect to the balancing of values and evidence. Given that values should *enrich*, not *replace* evidence that is to support decision-making under uncertainty, how should collaborative analyses proceed in order to secure both, deliberation and sound science?

The in-depth analysis of one specific example of public debate as a means of participatory governance in the UK (*GM Nation?*) leads to careful assessments paying due heed to the pitfalls outlined above (Joss, this volume): Alongside positive evaluations regarding the organization of a complex deliberative procedure, there was a specifically negative result concerning a noticeable disjuncture between policy and politics to be found: While the policy-makers were more interested in being provided with 'public opinion,' the public did not want to be restricted to this function. It wished to act as a politically and socially engaged participant in its own right and thus to influence the politics concerned with GM-crops. This disjuncture was reflected in its impact on policy-making: Although embedded in formal policy-making, the government's commitment toward the initiative was non-binding.

However, for some authors, strong governance, including strong elected bodies, *is* reconcilable with a high level of public participation, provided that participatory methods would shift so as to allow for 'real learning' (Hisschemöller, this volume). First, they would have to shift from their orientation toward finding a solution to finding the right problem. This shift would relate to the articulation and testing of rival hypotheses through involving knowledge from a variety of sources. Second, stakeholder dialogue on conflicting arguments should be promoted rather than prevented, thereby taking stakeholders as interested persons and groups who, owing to their biased position, have specific knowledge to offer.

Given these critical, albeit constructive evaluations, the fact that all institutions of extended expertise represent tricky remedies against technocracy as they, too, are compounded by a number of trade-offs should not be underestimated (Radaelli 2002).

The first such trade-off links democracy and time. In some of the most sensitive regulatory policies such as risk management, efficient and effective decision-making may rule out extensive consultation, public participation, and democratic audits of expertise in order to secure a competitive advantage. There are only a few voices that downplay this problem: In their view, deliberation, as an early warning system of public concerns, might actually facilitate decisions rather than merely delay them – not least due to the fact that they may prevent from later costly lawsuits (Hamlett 2003: 130).

The second trade-off is between democratization and the 'level' at which policy choices are made. Most of the instruments used to democratize and legitimize expertise, such as consensus conferences, citizens' juries, and deliberative polls, perform particularly well at the local level. Yet, how can participatory instruments be transferred from the local level to the national or the European level? Although some suggestions have been made (Cohen and Sabel 1997; Klüwer 2000), this remains, as yet, an open question. Research agendas for evaluating and improving public-participation exercises have only just been proposed (Rowe and Frewer 2004).

The third trade-off is between simplification and participation. One of the main trajectories of regulatory reform is the simplification of the policy process. In its annual reports on 'better law-making', the European Commission has made simplification of the policy process one of its top priorities. While simplicity is not a goal in itself, it is one of the best guarantees in terms of transparency and accountability. Additional regulation of expertise can act as a vehicle for opaque decisions and lack of clear points of responsibility and accountability. This calls for a coordination of traditional liberal-democratic institutions such as courts and legislatures with the activities of deliberative bodies, thereby making use of each institution's specific functions (Hamlett 2003: 132).

Regulation of (extended) expertise should therefore be clearly linked to policy learning: How do guidelines, registers, participatory mechanisms and other tools of democratization of expertise feed into the policy process, specifically given the two main governance gaps, namely the operational one and the participatory one. The operational gap occurs whenever policymakers and public institutions find themselves lacking information, knowledge and tools they need to respond to the daunting complexity of policy issues. The participatory gap indicates the "difficulty for a common understanding of, and therefore agreement on, critical policy issues. This has sometimes led policymakers to exclude the general public or particular stakeholders from their deliberations" (Reinecke and Deng 2000).

To overcome both types of gaps, procedural solutions have been proposed. Regardless of the origin of warning (civil society, science, politics), politicians ought to begin the decision-making process by consultation and democratic debate appropriate to the stakes involved – this, however, should happen in a coordinated fashion. To this end, Weill suggests the widespread use of contracts between experts and sponsors, contracts that are publicly available. Contracts set "the conditions and procedures for the panel and for determining in advance questions such as liability, availability of information, and ethics" (Weill 2003: 202). Analogously, in its report on democratizing expertise the EU (EU-Commission 2001) has recommended "five action lines for enhancing the credibility and effectiveness of expertise in the European Commission policy-making process. Four of these action lines related to specific aspects of mobilizing and interacting with experts: the development of an inventory; use of participatory procedures; broadening the expert base; and improving risk governance. The fifth concerned the development of a set of Commission guidelines, which, in effect, would provide a mechanism for implementing the other four" (Cross 2003: 190). The instruments are, first, guidelines on the collection and use of expert advice in the Commission that, over time, could form the basis for a common approach for all institutions and member states. Second, a 'checklist' has been appended, helping departments to design the most appropriate way of approach to experts according to specific circumstances (e.g., when and how to implement stakeholder consultation).

However, despite or because of such regulations, guidelines and checklists, the very legitimacy of participatory policy instruments can easily deteriorate if participation is an end in itself, leaving it unclear as to how it contributes to the actual problem solving. Again, several questions arise, e.g., concerning the participants, the process as well as the desired outcome of participatory policy instruments. First, the extent to which citizens *care to be involved* in decision-making might be questioned. The example of the United States reveals an interesting phenomenon: Although legal structures would allow for a high degree of participation and democratized decision making, this potential is not necessarily used by members of the general citizenry (Lahsen 2005: 159f). Second, although participation is recognized at the management stage, it is yet to be accepted at the knowledge-creation stages. Di Marchi calls for "constructing new methods of decision-making", for "the full realisation of a kind of participative governance is enormously difficult, as it requires broad changes in terms of professional and institutional practices, and the implementation of new tools and procedures for information sharing, concertation and deliberation" (de Marchi 2003: 175). Third, the question "participation for what?" should not go unnoticed. "There is a potential clash between greater inclusion and postponement of decisions, which are urgent and important. In some cases, participation might even be advocated instrumentally to defer certain types of decision. For example, promoters of new technologies (in the life science or other fields) can largely profit from a legislative vacuum derived from the delays in arriving at agreed upon regulations" (de Marchi 2003: 175).

No wonder, then, that several authors have recently identified a series of institutional forms of expertise testifying to "a broader swing back toward a technocratic model of governance in the United States. Expressions of this shift include: a rise during the past decade in official discourses on 'risk assessment,' 'sound science,' 'evidence-based decision-making;' a retreat from precautionary approaches to regulation; an attempt to cut back on citizen participation in environmental decisions ...; and, in the court system a partial displacement of jury trials by judicial pre-screening of scientific and technical evidence" (Jasanoff 2003: 158). We agree with Jasanoff and Nowotny, who hold that this push toward expert systems, benchmarking and evidence-based policy pursues a direction different from that exemplified by efforts to 'democratize expertise' through participatory models. "At best, these expert systems might become technologies of pluralizing expertise. They will contain a strong element of evidence-based experience and expertise that goes with it" (Nowotny 2003: 155). However, looking at these novel solutions at the interface of scientific expertise and political decision-making, we have to admit that they *are* numerous and that they cover the whole continuum from highly inclusive and highly coordinated deliberative processes on various 'levels' of decision making to highly exclusive and highly regulatory, evidence-based forms.

It seems that none of these forms is in itself a guarantee for bringing about 'better knowledge' for 'better politics.' Neither the sheer multiplicity of forms, ranging from participatory to regulatory, a bad thing. In our view, the multiplicity of forms rather depicts the complexity of responses to a complex problem. The upshot of our argument is that technocracy will not simply disappear, exactly because scientific expertise is operating in an increasingly politicized environment. Participation, accountability, due consultation, and transparency of decisions are part and parcel of the ideal-type of good governance, be it through citizen's juries, regulatory agencies or any combination of similar institutions (Majone 1996: chapter 13). In our view, to aver that there is only one choice, namely between technocratic and participatory solutions, is wrong. Accepting a systems theoretical approach, according to which science and politics operate along different primary codes (truth/power) we do not necessarily have to *accept,* but to *expect,* the emergence of a variety of forms that operate in-between and across participatory and regulatory modes. Consequently, we find various efforts at coordinating regulatory and participatory elements as well as regulating participatory elements (see above).

MAKING A CASE FOR EXTENDED EXPERTISE AS EXERCISES IN CIVIC RESPONSIBILITY

The multiplicity of forms and forums in which (scientific) expertise and (political) decision-making takes place should not obscure the bases they all rest on. Ultimately, they all rely on and, in actual fact, contribute to bringing about, the responsible citizen. Scientists, politicians and lay experts are united by being addressed in two capacities: as experts *and* as citizens.

To begin with, intermediary institutions designed to scrutinize knowledge for politics call for expertise that is not to be equated with scientific knowledge. Rather, as to the type of knowledge, expertise belongs neither to science nor to politics but is a hybrid that contains scientific as well as 'other' components. Those 'other' components help to contextualize scientific knowledge, thereby directing the knowledge away from specialized views and partial interests toward the common good. In fact, the ideal of a 'strong democracy' (Barber) in which institutions of extended expertise are embedded, calls for no less than this. Apart from political control, modern societies increasingly rely on societal self-control, civil society and the personal responsibility of their citizens.

This reorientation does not only apply to scientific experts but to other members of a participatory setting as well. The lay expert, in particular, is addressed as 'citoyen' taking the role of a citizen for whom the common good takes center stage when, e.g., deliberating the desirability of certain technical developments (Skorupinski and Ott 2002: 9). In a similar vein, Jasanoff looks at expertise as a form of 'delegated authority' in need of careful vigilance on the part of the lay experts. In this arrangement, democratic publics "only grant to experts a carefully circumscribed power to speak for them in matters requiring specialized judgment… Whether through direct participation or through organized questioning, the public has both *a right* and *a duty* to ask experts and their governmental sponsors whether appropriate knowledge is being deployed in the service of desired ends" (Jasanoff 2003: 159, italics added, SM and PW). This "mini-republic of ideas" includes experts who "impose on each other a degree of critical peer scrutiny that society can ill afford to do without" (Jasanoff 2003: 161). In addition to this, however, procedures designed to gain external accountability are emerging. In these settings, experts – lay, professional or scientific – are considered as members of the general citizenry. In this capacity, they have to synthesize their respective knowledge and stakes according to the Common Good (Maasen and Kaiser, forthcoming; Jasanoff, this volume).

Recent research is cautiously optimistic: Following these findings, public participation and support *can* produce more policy effectiveness (van de Peppel, quoted by Bressers and Rosenbaum 2000); conflict *can* be a resource in the policy process in terms of learning and social-institutional innovation (Dente et al. 1998); and the public *can* go beyond narrow interests and accept a socially inclusive view of risk management (Halfacre et al. 2000). While we would not deny this entirely, we do emphasize the series of shifts on which these findings and their fairly optimistic outlooks rest: First, we see a shift in the institutional settings which tend to become more 'hybrid'; second we discern a novel use of scientific knowledge, which, as expertise, is transgressive by necessity; thirdly, we detect a realignment of experts who are addressed as citizens transgressing their individual knowledge, values, and interests.

In this sense, participatory settings can be called 'agora,' a domain in its own right, yet not a domain of *primary* (Nowotny 2003: 156) but, as we see it, of *secondary* knowledge production. In this intermediary domain – neither purely scientific nor purely political – knowledge of various sources, as well as competing values and interests, can be discussed and negotiated. Here, political positions can be developed as a result of joint expertise and deliberatively produced policy recommendations. Given a certain media attention, input from participatory settings cannot easily be ignored by formal political bodies. In fact, publicity accomplishes a specific „pressure toward reflexivity", that both irritates and enriches formal power-driven policymaking (Neidhardt 1996: 66). More particularly, according to Matthias Kettner, this is what provides deliberative settings with "*contextual democratic legitimation,...* irrespective of the fact how (un)democratic their internal structures of membership and decision-making may be" (Kettner 2000: 404, italics added, SM and PW). While not directly influencing, participatory settings still *resonate* with the political subsystem (Sclove 1994; Edwards 1999).

Further, it should not go unnoticed that the more participatory variants of knowledge-based decision-making are arrangements that fundamentally rely on their being *exercises*. They are careful stagings of deliberative reasoning that manage to reduce, albeit as 'special events,' three types of complexity: The *factual* complexity is reduced by coordinated procedures of transferring, discussing and deciding about knowledge; the *temporal* complexity is reduced by decision 'until further notice' (such decisions will return on the agenda, a required by the actual state of knowledge or values); the *social* complexity of the interaction of heterogeneous actors (politicians, scientists, lay experts) is reduced by recourse to the participants' civil competences. This staged clearness allows for temporary understandings and compromises on a case-by-case basis. It rests on a basic resource that it co-produces in the course of the exercise: responsibility.

Following those who proclaim the 'end of the social' or the 'era of raging subjectivity,' respectively, it is hard to explain what it is that provides appeals to responsibility with such plausibility. Are our highly individualistic societies not antithetic to civic activities, in principle? From a 'governmentalistic' point of view (Foucault 2000), however, one arrives at the opposite conclusion: Modern political governance makes use of everybody's capacities to conduct themselves and others. In this view, neo-liberal societies fundamentally rely on techniques of governmentality. Such technologies consist of "mundane programs, calculations, techniques, apparatuses, documents and procedures through which authorities seek to embody and give effect to governmental ambitions" (Rose and Miller 1992: 175). Among those techniques and procedures we find participatory technology assessments, citizens' juries, transdisciplinary modes of knowledge production, all said "to enhance social cohesion and strengthen civic discourse" (Belluci et al. 2002: 278) by way of skillfully combining output-oriented policy analysis aspects with process oriented discourse aspects (ibid.). Participation, wherever it occurs, thus becomes a "technology of citizenship ... by which government works *through* rather than *against* the subjectivities of citizens" (Cruikshank 1999: 69, italics added, SM and PW).

Enacting citizenship, however, is not restricted to lay people but rather embraces all participants in such exercises: politicians, administrators, lawyers, scientists, and so on. The political rationality underlying responsible participation in knowledge-based decision-making has been called 'ethopolitics,' "which works through the values, beliefs, and sentiments thought to underpin the techniques of responsible self-government and the management of one's obligation to others" (Rose 2002: 1399). Appeals to responsibility, hence, gain their acceptability by way of the ethopolitical rationality of modern neoliberal societies. By the same token, participation appears to be a procedure by which modes of responsible self-government become a political technology that not only complements other forms of techno-political decision-making, e.g., regulatory ones, but, at the same time, affords its most important resource: the responsibly acting political subject (Sutter 2005).

This development is a highly ambivalent one, however: the increase in autonomy (the *possibility* to know and decide) is inevitably accompanied by an increase in heteronomy (the *need* to know and decide) in ever-more science-based policy institutions, on ever-more issues, in more or less rigid procedures (Weber [1920]1993). The involvement of subjects and their capacity to commit themselves to responsible decisions, is a double-edged sword. While the emergence of inclusive forms of knowledge-based decision-making certainly advances democratic values of participation in societal decisions under uncertainty, it also advances responsibilization (O'Malley 1996), that is, individualization of societal risk-taking (Lemke et al. 2000).

We began this essay by stating that reopening the case of scientific advice to politics means to find novel solutions to the dark side of expertise: technocracy. We conclude this essay with a caveat: may extended peer review not lead to extended technocracy by way of skillfully putting citizens into service of solving the irresolvable dilemmas at the interface of science and politics.

* *Universität Basel, Switzerland*
***Universität Bielefeld, Germany*

NOTES

[1] Generally, technocrats are considered to be promoting social order, economic growth, national security and most of all, they are said to be more interested in technological progress and material productivity than in distributive questions about social justice (Brint 1990: 366), let alone sustainability, ethics and safety – issues that have come to the fore during the last fifteen years. These latter demands, in particular, are said to call for more than technocratic knowledge oriented toward 'managing' governmental activity. They should be substituted or complemented by deliberative practices expressing broader societal interests and ideals, instead

[2] The examples for the use of 'blurring' as a descriptive concept are countless. The most pertinent theoretical approach is the constructivist claim that there is no meaningful difference between scientific and everyday knowledge (Knorr-Cetina 1981).

REFERENCES

Abels, G. (2003), 'Experts, Citizens, and Eurocrats – Towards a policy shift in the governance of biopolitics in the EU', *Europe and Integration online Papers (EIoP)* **6/19**, http://www.eiop.or.at/eiop/texte/2002-019a.htm.

Abels, G. and A. Bora (2004), *Demokratische Technikbewertung*, Bielefeld: transcript-Verlag.

Bimber, B. (1996), *The Politics of Expert Advice in Congress*, New York: State University of New York Press. http://www.sscf.ucsb.edu/~survey1/poltran2.htm.

Bellucci, S. et al. (2002), 'Conclusions and recommendations', in S. Joss and S. Bellucci (eds.), *Participatory Technology Assessment. European Perspectives*, London: Centre for the Study of Democracy, pp. 276–87.

Brint, S. (1990), 'Rethinking the policy influence of experts: From general characterizations to analysis of variation', *Sociological Forum* **5**, 3: 361–85.

Cohen, J. and C. Sabel (1997), 'Directly-deliberative polyarchy', *European Law Journal* **3**, 4: 313–42.

Cross, A. (2003), 'Guidelines for expert advice. Drawing up guidelines for the collection and use of expert advice: The experience of the European Commission', *Science and Public Policy* **30**, 3: 189–92.

Cruikshank, B. (1999), *The Will to Empower. Democratic Citizens and other Subjects*, Ithaca, London: Cornell University Press.

Dente, B., P. Fareri and J. Ligteringen (1998), *The Waste and the Backyard. The Creation of Waste Facilities: Success Stories in Six European Countries*, Dordrecht, Boston and London: Kluwer Academic Publishers.

Després, C., N. Brais and S. Avellan (2004), 'Collaborative planning for retrofitting suburbs: Transdisciplinarity and intersubjectivity in action', *Futures* **36**: 471–86.

Edwards, A. (1999), 'Scientific expertise in policy-making: The intermediary role of the public sphere', *Science and Public Policy* **26**, 3: 163–70.

Ellul, J. (1964), *La technique ou l'enjeu du siècle*, Paris: Armand Colin.

EU-Commission (2000), *Science, Society and the Citizen in Europe*, Working Document.

EU-Commission (2001), Europena Governance, a White Paper, archived at http://www.europa.eu.int/comm/governance/white_paper/index_en.htm.

Foucault, M. (2000), 'Die Gouvernementalität', in U. Bröckling, S. Krasmann, T. Lemke (eds.), *Gouvernementalität der Gegenwart. Studien zur Ökonomisierung des Sozialen*, Frankfurt a.M.: Suhrkamp, pp. 41–67.

Gibbons, M., C. Limoges, H. Nowotny, S. Schwartzman, P. Scott, M. Trow (1994), *The New Production of Knowledge*, London: Sage Publications.

Gilpin, R. and R. Wright (eds.) (1965), *Scientists and National Policy Making*, New York: Columbia University Press.

Guston, D.H. and K. Kenniston (1994), *The Fragile Contract*, Cambridge: MIT Press.

Habermas, J. (1962), *Strukturwandel der Öffentlichkeit*, Neuwied: Luchterhand.

Halfacre, A.C., A. Matheny and W. Rosenbaum (2000), 'Regulating contested local hazards: Is constructive dialogue possible among participants in community risk management?', *Policy Studies Journal* **28**, 3: 648–67.

Hamlett, P.W. (2003, January), 'Technology, theory and deliberative democracy', *Science Technology & Human Values* **28**: 112–40.

Jasanoff, S. (2003), '(No) Accounting for expertise', *Science and Public Policy* **30**, 3: 157–62.

Kettner, M. (ed.) (2000), 'Welchen normativen Rahmen braucht die Angewandte Ethik?', *Angewandte Ethik als Politikum*, Frankfurt a.M.: Suhrkamp, pp. 388–407.

Klüwer, L. (2000), 'European-wide participation and the use of the precautionary principle', *Danish Board of Technology*, December, 2000.

Knorr-Cetina, K. (1981), *The Manufacture of Knowledge*, Oxford et al.: Pergamon Press.

Krücken, G. and P. Hiller (eds.) (1997), *Risiko und Regulierung. Soziologische Beiträge zur Technikkontrolle und präventiver Umweltpolitik*, Frankfurt a.M.: Suhrkamp.

Lahsen, M. (2005), 'Technocracy, democracy, and U.S. climate politics: The need for demarcations', *Science, Technology & Human Values* **30**: 137–69.

Lakoff, S.A. (1966), *Knowledge and Power*, New York: Free Press.

Lapp, R.E. (1965), *The New Priesthood*, New York: Harper & Row.

Lemke, T., S. Krasmann and U. Bröckling (2000), "Gouvernementalität, Neoliberalismus und Selbsttechnologien. Eine Einleitung", in U. Bröckling, S. Krasmann and T. Lemke (eds.), *Gouvernementalität der Gegenwart. Studien zur Ökonomisierung des Sozialen*, Frankfurt a.M.: Suhrkamp, pp. 7–40.

Maasen, S. and M. Kaiser (forthcoming), "Vertrauen ist gut. Verantwortung ist besser. Wissenschaft und Politik in der Vertrauenskrise: Weiter mit Verantwortung?", to appear in C. Rehmann-Sutter, J.L. Scully, R. Porz and M. Zimmermann-Aklin (eds.), *Gekauftes Gewissen – Die Rolle der Bioethik in Institutionen*, http://pages.unibas.ch/wissen/main/forschung/veroeffentlichungen.htm.

Majone, G.D. (1996), *Regulating Europe*, London: Routledge.

De Marchi, B. (2003), 'Public participation and risk governance', *Science and Public Policy* 30, 3: 171–76.

Marcuse, H. (1961), "Das Problem des sozialen Wandels in der technologischen Gesellschaft", in P.E. Jansen (ed.) (1999), *Herbert Marcuse Nachgelassene Schriften 1. Das Schicksal der bürgerlichen Demokratie*, Lüneburg: Zu Klampen, pp. 37–66.

Meynaud, J. (1969), *Technocracy*, New York: Free Press.

Murswieck, A. (1994), 'Wissenschaftliche Beratung im Regierungsprozess', in A. Murswiek (ed.), *Regieren und Politikberatung*, Opladen: Leske + Burdrich, pp. 103–19.

Neidhardt, F.M. (1996), 'Öffentliche Diskussion und politische Entscheidung. Der deutsche Abtreibungskonflikt 1970–1994', in W. van den Daele and F. Neidhardt (eds.), *Kommunikation und Entscheidung: Politische Funktionen öffentlicher Meinungsbildung und diskursiver Verfahren*, Berlin: Edition Sigma, pp. 53–82.

Nowotny, H. (2003), 'Democratizing expertise and socially robust knowledge', *Science and Public Policy* 30, 3: 151–56.

Nowotny, H., P. Scott and M. Gibbons (eds.) (2001), *Re-Thinking Science: Knowledge and the Public in an Age of Uncertainty*, London: Polity Press.

O'Malley, P. (1996), 'Risk and responsibility', in A. Barry, T. Osborne, and N. Rose (eds.), *Foucault and Political Reason*, Chicago: University of Chicago Press, pp. 189–207.

Van de Peppel, R. (2000), cited in Bressers, H. and W.A. Rosenbaum (2000), 'Innovation, learning, and environmental policy. Overcoming a plague of uncertainties', *Policy Studies Journal* 28, 3: 523–39.

Price, D.K. (1967), *The Scientific Estate*, Cambridge: The Belknap Press of Harvard University Press.

Radaelli, C. (2002), 'Democratizing expertise?', in J.R. Grote and B. Gbikpi (eds.), *Participatory Governance. Political and Societal Implications*, Opladen: Leske + Budrich, pp. 197–212. Available online at: http://www.bradford.ac.uk/acad/ssis/staff_contact/radaelli/demexp.html.

Reinecke, W.H. and F.M. Deng (2000), *Critical Choices. The United Nations, Networks, and the Future of Global Governance (IDRC, Ottawa)*, http://www.idrc.ca/books/921.

Rose, N. (2002), 'Community, Citizenship, and the Third Way', *American Behavioral Scientist* 43, 9: 1395–411.

Rose, N. and P. Miller (1992), 'Political power beyond the state', *British Journal of Sociology* 43, 2: 173–205.

Rowe, G. and L.J. Frewer (2004), 'Evaluating public-participation exercises: A research agenda', *Science, Technology & Human Values* 29: 512–56.

Saretzki, U. (1997), 'Demokratisierung von Expertise? Zur politischen Dynamik der Wissensgesellschaft', in A. von Klein and R. Schmalz-Bruns (eds.), *Politische Beteiligung und Bürgerengagement in Deutschland*, Baden-Baden: Nomos.

Schelsky, H. (1965), 'Der Mensch in der wissenschaftlichen Zivilisation', in H. Schelsky, *Auf der Suche nach Wirklichkeit, Gesammelte Aufsätze*, Düsseldorf: Eugen Diederichs, pp. 439–80.

Sclove, R.E. (1994), 'Citizen-based technology assessment? An update on consensus conferences in Europe', http://www.manymedia.com/loka/loka.1.12.txt.

Skorupinski, B. und K. Ott, (2002), 'Partizipative Technikfolgenabschätzung als ethisches Erfordernis. Warum das Urteil der Bürger/innen unverzichtbar ist', edited by TA-SWISS, http://www.ta-swiss.ch/www-remain/reports_archive/publications/2002/DT31_Bericht_kompl.pdf.

Stone, D.A. and M.D. Garnett (eds.) (1998), *Think Tanks Across Nations. A Comparative Approach*, Manchester: Manchester University Press.

Sutter, B. (2005), 'Von Laien und guten Bürgern: Partizipation als politische Technologie', in A. Bogner and H. Torgersen (eds.), *Wozu Experten? Wissenschaft und Politik: Sozialwissenschaftliche Diagnosen einer Beziehung im Umbruch*, Wiesbaden: Verlag für Sozialwissenschaften (in print).

Touraine, A. (1971), *The Post-Industrial Society*, New York: Basic Books.

Weber, M. ([1920] 1993), *Die protestantische Ethik und der "Geist" des Kapitalismus* (Textausgabe auf der Grundlage der ersten Fassung von 1904/05 mit einem Verzeichnis der wichtigsten Zusätze und Veränderungen aus der zweiten Fassung von 1920), Bodenheim: Athenäum Hain Hanstein.

Weill, C. (2003), 'Can consultation of both experts and the public help developing public policy? Some aspects of the debate in France', *Science and Public Policy* **30**, 3: 199–203.

Weingart, P. (2001), *Die Stunde der Wahrheit? Zum Verhältnis der Wissenschaft zu Politik, Wirtschaft und den Medien in der Wissensgesellschaft*, Weilerswist: Velbrück Wissenschaft.

Weingart, P. (2003), 'Experte ist jeder, alle sind Laien', *Gegenworte* **11**: 58–61.

Weingart, P., A. Engels und P. Pansegrau (2002), *Von der Hypothese zur Katastrophe. Der anthropogene Klimawandel im Diskurs zwischen Wissenschaft, Politik und Massenmedien*, Opladen: Leske + Budrich.

Zhou, Z. (2002), *Risikomanagement durch Systemverzahnung. Umweltqualitätsnormung zwischen Wissenschaft und Recht*, Wiesbaden: Deutscher Universitätsverlag.

CHAPTER 2

ALEXANDER BOGNER[*] AND WOLFGANG MENZ[**]

BIOETHICAL CONTROVERSIES AND POLICY ADVICE: THE PRODUCTION OF ETHICAL EXPERTISE AND ITS ROLE IN THE SUBSTANTIATION OF POLITICAL DECISION-MAKING

At the beginning of 2002 the German parliament took a decision on the permissibility of embryonic research. The compromise reached had neither the compelling logic of the liberal position nor the moral consistency of the opponents of research involving the destruction of human embryos: it allows research on *imported* embryonic stem cells which originated before January 2002. The decision was preceded by a public discussion in talk shows and newspapers where, for a long period before the fundamental political decision, the most important arguments and positions on the question of the ethical legitimacy of stem cell research were debated. At the end of November 2001 the recommendations of the ethics councils were made available. The *Nationaler Ethikrat* (National Ethics Council) and the *Enquete-Kommission 'Recht und Ethik in der modernen Medizin'* (Study Commission on Law and Ethics in Modern Medicine) expressed the anticipated dissent in the commissions by formulating divergent positions and documenting them in separate votes. A similar situation occurred a little later in Austria. At about the same time as the German National Ethics Council was being established, the Austrian chancellor convened a bioethics commission which drew up a statement on stem cell research. As in the German case, competing positions were expressed and documented.

On the basis of this example of 'ethical assistance' to political decision-makers, we discuss in this chapter the following questions: Can one identify a social meaning of the advice provided by expert commissions under conditions of the absence of clarity? Or, more generally: What does the "new institutionalisation of morality" (Kuhlmann 2002) mean for the relationship between expertise and politics? And what follows from the disagreement of the commissions' experts for the legitimation of political decision-making?

Our chapter, dealing with this "new complexity in the relationship between science and politics" (Weingart 2001a: 80), is structured in two sections. In the first section, against the background of the sociological tradition, we critically address the basic assumptions of the theory of reflexive modernisation concerning the role of expert knowledge in the face of new risks. In the second section we outline, in the

---

Sabine Maasen and Peter Weingart (eds.), *Democratization of Expertise? Exploring Novel Forms of Scientific Advice in Political Decision-Making – Sociology of the Sciences, vol. 24*, 21–40.
© Springer Science+Business Media B.V. 2009

22   ALEXANDER BOGNER AND WOLFGANG MENZ

form of a series of theses, some findings of our qualitative interviews with members of the Austrian Bioethics Commission.

INSTITUTIONALISED DISAGREEMENT AND POLITICAL DECISIONS: CRITICAL REMARKS ON THE THEORY OF REFLEXIVE MODERNISATION[1]

Max Weber discusses the question of expert knowledge and political decision-making in terms of the relationship between the specialised German civil service (*Berufsbeamtentum*) and political leaders (Weber 1958). He is concerned that political action is subjected to an administrative-bureaucratic logic of its own, paralysing political initiative and influence. Although the 'leading spirit' of the politician depends on the technical knowledge of administrative experts, he also has to free himself from the administrative apparatus in order to make fateful decisions. How can the weights of the civil service, parliament and political leaders be balanced without either letting the bureaucracy inhibit the capability to decide or leaving unused the advantages of rational and specialised organisation?

In contrast to Weber, who demands the indispensable sovereignty of politics as opposed to the inescapable process of bureaucratisation, *Helmut Schelsky* sees the nature and task of politics in the functionality of technology. The state becomes a "technical state," where politics only has a fictitious capacity to take decisions (Schelsky 1965: 457). Weber's political decisions, as questions of fate, became bare matter-of-fact choices between different recommendations and expert opinions. Given the premise that the uncertainty of decision vanishes in the course of technological development, and that the best option according to scientific rules can be realised, politics becomes the executive body of expert reason.

In contrast to the above two authors, *Ulrich Beck* places the growing ambiguity, plurality and contradictoriness of expert knowledge at the centre of his analysis (Beck 1993, Beck and Bonß 2001). When new uncertainties appear and are recognised, the traditional-modern rationality, which is orientated towards efficiency and productivity, is split up into different, competing and contradictory rationalities. In the course of modernisation, side-effects are produced and spaces of non-knowledge are revealed which question modernisation itself. The controversial interpretations of (knowledge-based) risks reveal systematic insufficiencies of expert knowledge and prove that the expert's promises of security are exaggerated. Expert knowledge becomes both the stimulus and the medium of a struggle about the power to define. The quarrel between experts and counter-experts becomes part of the enlightenment of modernity about itself. In these conditions, argues Beck, political decisions cannot simply be taken at the behest of the experts and without public discussion and participation.

Today, hardly anyone would share the optimistic view that the side-effects and the recognition of non-knowledge will lead to the desired revision of the path of modernisation. We can observe a political-institutional pacification of previously central risk conflicts. Beck's counter-experts grew up, became institutionalised, and have been funded by public money and have developed their own modes of selection and hierarchisation. As the established interpreters of the deficits of modernisation, they are part of the dissonant background choir of modernisation. The struggle over the power to define has become a pluralistic debate.

In addition, Beck's optimism conceals some systematic reasons which imply a certain danger of misjudging the new political meaning of pluralism and dissent. Beck's theory of expert knowledge and modernisation is based on an ambiguous concept of risk, seeing it as a social construction but at the same time as a real threat independent of the observer or the discourse. This objectivist reading of risk as an attribute of technology has been criticised by Luhmann (1991) and other authors who also refer to a constructivist epistemology (Japp 1996, Tacke 2000). Luhmann argues that the concept of risk should not be reserved for the – albeit spectacular – potentials of danger and threat. In the concept of risk, different elements of the various processes of de-traditionalisation since early modernity intersect. In the course of the dissolution of traditional beliefs about fate and predestination, the future is no longer constructed as an extended present and new forms of coping with this new future, open to decision, have to be found. The immense significance of risk – in the sense of decision-making under the "pressure of contingency" (Japp 1992: 38) – refers to the gradual strengthening of a specific modern type of rationality. The aim is to act purposefully, rationally and in a calculating way in relation to the future, in order to realise new chances of action (Bonß 1995). In this respect risk has to be seen as a specifically modern case of an unrealisable imposition of rationality: we have to decide without knowing the 'right' solution or, to be more precise, with the certainty that there is no single, optimal solution.

This insight also leads us to a more critical assessment of the political significance of expert disagreement. If we cannot regard the decision about a (fundamentally) uncertain future as a result of superior knowledge and systematic scientific rationality, the decision comes to depend on symbolism and strategies. The traditional rationality of political decision-making is replaced by a kind of *politics of rationality*. Politics has to solve the problem of reconciling the societally influential idea of rationality with what has been seen of its way of dealing with uncertainty. Today, therefore, it is important for political decision-makers to make the public accept (on the basis of the formal granting of certain expectations concerning the process of the shaping of opinion) that *it is possible to decide and decisions must be taken*. One way of establishing a credible and convincing moment of decision is precisely by establishing a role for an institutionalised (counter-)expertise.

We have been able to observe this need to create a moment of decision in the course of the process of decision-making on the question of the permissibility of stem cell research. Although the statements of the expert commissions in Germany and Austria did contain hardly any new arguments as compared to the broad public debate, it was not considered acceptable for politicians to anticipate the experts' conclusions. The recommendations represented a decree that a *political* decision was now on the agenda. A political decision is legitimised only if it is assumed that it could have turned out differently. In short, politics in a sense *relies* on dissent. Although the commissions made clear the relevance of the topic and showed the possibility and necessity of regulation, they gave contradictory recommendations.

From the perspective of the political system this is quite functional: through the divergence of the expert's options, *politics as decision* actually becomes visible. If politics were just the execution of technical formulas or inherent necessities (as in Schelsky's concept), it would eventually disappear.

At the same time a transformation of the rationality of political decision-making takes place. In the German example, the decision about the permissibility of stem cell research was delegated to the 'sense of responsibility' or the 'conscience' of the members of parliament. In a similar way, in Austria the decision was handed over to the personal assessment of the minister who had to represent Austria's position in Brussels on the question of whether stem cell research, in the context of the EU's sixth framework programme, should be funded by taxation. In a sphere which is usually understood as an arena of superior reason, a scientifically supported and legitimised emotionality became the political rationality. Indeed, the arguments of the scientists and experts in the ethics commissions provide the necessary background knowledge and the patterns of legitimation. But the real decision takes place on a different level.

This does not mean that politics decides irrationally (in a value sphere outside expertise), nor that technical constraints determine politics. Hence, the relationship between science and politics cannot be represented adequately in the form of a model of dominance as described by Weber and – with the opposite normative assessment – Schelsky. Rather, our interpretation suggests that we should understand this relationship as a form of mutual instrumentalisation. For politics, the dissonant expert discourse is a prerequisite for a successful maintenance of the ability to act. For science uncertainty and pluralism become characteristics of quality, especially concerning its role as a legitimising resource for politics.

From this perspective the dissent of the experts cannot be seen as a correction potential of modernisation. On the contrary: the current controversies about bioethics show in an exemplary way that the interaction between expertise and politics is a fundamental precondition for the establishment of the political capacity to act via the affirmation of the existing logics of procedure. Conflicts among experts not only express the limits of the traditional-modern way of dealing with new uncertainties; they also provide a way of coping with such uncertainties.

THE PRODUCTION OF EXPERT DISAGREEMENT – THE CASE OF THE AUSTRIAN BIOETHICS COMMISSION

In this section we will substantiate our conceptual considerations with some insights derived from an empirical perspective. We take a closer look at the Austrian Bioethics Commission's internal processes of decision-making, the political role of the commission's work and its recommendations for public policy. We concentrate mainly on the question of how the discussion in a group of experts results in manageable decisions and political recommendations.

For our analytical perspective it is decisive that we do not consider the process of generating an expert vote mainly as a process of general understanding and agreement, in which a "peculiar non-coercive force of the better argument" (*"eigentümlich zwanglose(r) Zwang des besseren Argumentes"*) (Habermas 1971: 137, trans. by authors) can unfold itself. Rather, our analysis is inspired by concepts originating from strategic organisation analysis as formulated by Michel Crozier and Erhard

Friedberg (Crozier and Friedberg 1977, Friedberg 1993). This approach stresses that action within an organisation has to be understood in terms of the strategies and resources of action of concrete (primarily collective) actors in the context of organisational structures, rules and "games." Thus, we place the processes of negotiation and compromise-formation as well as the formal and informal structures of coordination at the centre of our analysis. Processes of decision-making cannot simply be explained by reference to a superordinate societal structure, but have to be understood in the context of the particular "local order" (Friedberg 1993). In a similar way, we draw on sociological analyses of micropolitics within organisational processes and decisions (Küpper and Felsch 1999, Küpper and Ortmann 1992, Ortmann et al. 1990).[2] From the perspective of political science, Nullmeier et al. (2003) formulate an ethnographic view of everyday action and interaction in political institutions below the level of macro-structures called 'micropolicy analysis,' which shares a similar perspective, in some respect, with organisational sociology. With respect to our subject, our thesis is that the formulation of the experts' votes cannot be understood as an uninfluenced 'free' shaping of opinion, nor a process determined by political macro-structures with an outcome determined beforehand. Rather, we postulate that it is an organised procedure of negotiating expertise which leads to the production of politically manageable dissent.

The Austrian Bioethics Commission was constituted in July 2001, one month after the German Chancellor Gerhard Schröder established the German National Ethics Council. The legal basis of the appointment of the Austrian commission was an order enacted by the Austrian Chancellor Wolfgang Schüssel. The commission has the task of advising the chancellor on bioethical questions and providing recommendations in relation to the challenges confronting the government as a result of biomedical progress. In the latter case, the commission has to deal with questions raised by the political system. Both the experts and the government single out two roles of the commission: first, policy advice; and second, the initiation and intensification of public discourse about bioethical issues. This means that the commission has to provide recommendations on current trends in the biomedical field, and also has to be a kind of early warning system, a discussion forum able to identify emerging or imaginable societal consequences of biomedical progress as a central theme. Furthermore, it is supposed to engage with an intensifying public debate. Even the commission members considered it a weakness that the commission did not set the agenda. But up to now the commission has acted in accordance with the day-to-day political agenda, to the extent that reactions to political inquiries have been central to its work. The commission's discussions have been aimed at developing recommendations and opinions on current problems (like stem cell research) and other biomedical issues which have been neglected in Austria, such as the ratification of the Convention on Human Rights and Biomedicine of the Council of Europe and the implementation of the Council Directive on the legal protection of biotechnological inventions.

The interdisciplinary Austrian Bioethics Commission consists of 19 experts, 15 men and four women (the German National Bioethics Council has 25 members). There are seven medical doctors (fields of activity: gynaecology, haematology, psychiatry, oncology and pathology), three human geneticists, two lawyers, two philosophers, two theologians, one sociologist, one computer scientist and one representative of the biotech industry (Novartis). In terms of different world views or 'ideologies,' the commission is distinguished by considerable heterogeneity: there is a Catholic theologian and an atheist, conservative medical doctors, and convinced social democrats. The members of the Austrian commission were appointed by Chancellor Schüssel, and the most important criterion of appointment was – according to the official statement – their specialised knowledge (Bundeskanzleramt 2001). But in fact nobody knows why exactly these 19 experts were appointed and who advised the chancellor. Even the members of the commission themselves cannot explain which criteria were decisive. The lack of transparency of the process has been repeatedly criticised, as has the fact that the commission was created as a pure expert council (Gottweis 2001).

The Austrian Bioethics Commission is therefore under considerable legitimation pressure in several respects. Criticisms from different organisations of disabled people culminated in the founding of a (counter-)commission, the *Ethikkommission FÜR die österreichische Bundesregierung* (Ethics Commission FOR the Austrian Federal Government, original emphasis), just a few months after the expert commission had been founded. Its name indicates its purpose – to be an advisory body for political decision-makers. It was explicitly set up to represent persons affected by the political decisions about bioethical matters of dispute, and not as a pure assembly of academic expertise like the 'official' bioethics commission. However, the pressure of legitimation did not only pertain to the process of appointment and the lack of representation of important social groups, but to the general question of how to institutionalise political advice in the form of expert commissions.[3] Thus, the alternative commission does not only see its task as providing 'better' or 'counter-balancing' expertise for the government, but also involving the broader public in the process of shaping a political opinion and making decisions.

Participatory forms of formulating political objectives do not have a strong tradition in Austria, in contrast to countries such as Denmark, where commentators see the most far-reaching efforts to involve the public in decisions on issues of bioethics (see Joss and Durant 1995). However, we can observe a change in attitudes or an attempt to broaden the debate in Austria. For example, most recently and for the first time, a *BürgerInnenkonferenz* (citizens' conference)[4] took place, with a thematic focus on genetic diagnostics.[5] Like other initiatives in Europe (see Joss and Torgersen 2002), its goals were twofold. On the one hand, it was intended to stimulate a broad public discussion. On the other hand, its proposals, votes and demands were designed to influence political and economic decision-makers. The *BürgerInnenkonferenz* was shaped according to the Danish concept of the consensus conference with its aim of democratising technology policy (Winner et al. 1997). In this way, the *BürgerInnenkonferenz* was designed to be a form of institutionalised advice, in a sense competing with any expert committee. It remains to be seen whether it will be able to do this in practice. Certainly, it is unlikely that this kind of participatory bioethical advice will develop into a general challenge to expert committees, and that it will gain broad influence in state politics. Possibly, Austrian politics will exploit the results of institutionalised public participation efforts as a source of more reliable information on public opinion, because they promise to deliver deeper and more complex and authentic views as compared to opinion polls.[6]

Ethical Decision-Making Goes Pragmatic: The Issue and Subject of the Stem Cell Debate

So far, we have concentrated on the composition of the Austrian Bioethics Commission and on a broad outline of the context of its institutionalisation. We now turn to the internal procedures designed to provide ethical expertise. Since all our remarks are related to the recommendations of the commission on the moral tenability of stem cell research, we shall begin by characterising different positions within the commission in more detail. This recommendation (entitled *Stem cell research in the context of the EU's sixth framework programme research*)[7] dealt with the controversial issue of whether or not Austria should finance stem cell research in the context of the EU's sixth framework programme (2002–2006). The resolution of the Bioethics Commission amounts to six pages, and the text is divided into two parts. The first part (two and a half pages long) presents a text that was unanimously agreed upon. In this part the experts emphasise that, from an ethical point of view, it is inadvisable to encourage exaggerated or premature expectations of a cure for any disease. In addition, the medical relevance and currently increasing importance of adult stem cells is emphasised and so the commission welcomes the decision of the Council of Ministers of the European Union to prioritise funding for research on adult stem cells. The Austrian commission also agrees unanimously with the European Parliament's decision to deny funding for reproductive cloning, the production of embryos for research purposes, and the modification of the genetic heritage of human beings.

This first part of the commission resolution is characterized by an expressly declared consensus within the group about the ethical difficulties arising from research with embryonic stem cells. It is followed by divergent judgements concerning the morality of doing research with embryonic stem cells, which are represented by the positions A and B. Position A supports funding research for embryonic stem cells, but recommends that strictly defined conditions have to be met, for example high-level peer review. In addition, only those stem cell lines are to be used which already existed before a given date (according to the corresponding regulation in Germany), so that the destruction of supernumerary embryos created via IVF treatment is not encouraged for purposes of stem cell research. Position B opposes the encouragement of stem cell research. Unintended consequences and the potential dangers resulting from the liberalisation of research using human embryos are emphasised. Following the well-known 'slippery slope' argument, it is claimed that in the end this encouragement will lead almost inevitably to liberalising embryonic research or even lead to the deliberate creation of embryos for research purposes. This trend would inevitably result in an increased societal acceptance of the availability of, and consequently the capacity for, the instrumentalisation of human life. Eleven members assented to Position A, while eight voted for the more sceptical position B.

How did such partly unanimous, partly divergent recommendations come into being? Which processes of negotiation, persuasion and compromise formation took place? What coalitions were formed? Which strategies were followed? What roles did different kinds of 'scientific knowledge' and divergent forms of expertise according to the respective disciplinary backgrounds play? And last but not least: can we consider ethical recommendations to be a form of scientific policy advice at all?

In the following we will present the preliminary results of our empirical study on the production of expertise by the Austrian Bioethics Commission. Our analysis is based on semi-structured face-to-face interviews with 18 of its 19 members. In a second step, we will turn to the relationship between the 'commissionary expertise' and the political reasoning and legitimising efforts in public argumentation. What does disagreement among the 19 experts entail for the substantiation and justification of the decision about whether or not stem cell research should be allowed? What role did the institutionalisation of bioethical expertise play for the system of political decision-making?

> *Thesis: In the discussion process, fundamentalist points of view are marginalised. From the very beginning we can see a strong 'pragmaticisation' of decision-making. Rather than deep reflection, producing results is the main consideration.*

Because they have to address concrete questions relevant to day-to-day politics, bioethics commissions hardly provide any room for fundamental debates about ethical questions and positions, and no room at all for deeper questions such as the conceptualisation of ethics as such. Therefore, the commission gets into a paradoxical situation where there is little clarification of its subject. This does not imply that the fundamental ethical contexts of the relevant positions no longer play a role. The interviewees often pointed to different ethical traditions linked to the different ways of arguing. However, such differences were kept at a low level during the discussions within the commission. In other words: an ethics commission does not function like a philosophical seminar, where participants discuss fundamental ethical considerations and the sustainability of different theoretical approaches (such as utilitarianism, deontological positions, etc.) with respect to certain biomedical questions. It is not the strength of the ethical argumentation that is decisive, and there is no debate on fundamentals. Rather, the question is whether a certain 'ethical basis' or orientation gets support from a majority. This does not imply that *un*convincing ethical positions will in general get support from a majority. Indeed, there is no serious assessment of how convincing divergent ethical positions are in practice – this is what we mean by 'pragmaticisation.' Additionally, a given position does not need to be founded in an explicit ethical argumentation. Rather, it is expected to result from a convergence with certain cultural *Leitbilder* (ideals of how to live, body images, etc.), which do not need to be present in a reflected way. From this perspective, an ethical argument supported by a majority does not imply a superior cognitive quality; rather, it shows a lucky convergence of societal dispositions and philosophical reflections. Last but not least, it would be unrealistic to expect an explicit ethical foundation for single positions to be provided by a bioethics commission primarily consisting of 'ethical lay persons' and working on a tight schedule.

Within the commission, a text on the issue of stem cells was proposed at an early stage. One member of the commission drew up this text on his own. The debate centered on this text, and it served as a trigger for discussions which ultimately led to a differentiation of positions. This is also reflected in the final paper. The more explicitly critical position B reads to a considerable extent like a comment on the other position.

Several members of the commission expressed their regret to us about the lack of depth of the ethical discussion. There are two sets of reasons for this 'pragmaticisation of the ethical decision-making process.' First, the commission was under pressure from the political sphere, because Austria had to take an early decision on where to position itself in the deliberations on the sixth framework programme of the European Union. However, it was not only external pressure from the minister responsible, Elisabeth Gehrer, or political influence that made it necessary to come to a swift decision; rather, this was (also) due to the genuine strategic self-interest of the commission. Had they delayed their statement by prolonging the debate, this would have resulted in a self-disempowerment of the commission. If the findings had been issued after the date relevant for the political decision, it would only have amounted to a critique or acclamation issued after the decision had been taken. The commission's own claim to advise the politicians would not have been substantiated, and the commission would have jeopardised its own function and performance.

Secondly, sacrificing the fundamental ethical debate was also significant for productivity within the commission. It made it possible to avoid potential conflicts, and it enabled the formation of coalitions between members otherwise supporting very different fundamental ethical positions. Excluding basic questions by pursuing ethical pragmatism increases the capacity to come to unanimous decisions among different positions.

The Politics of Ethics – Ethics as Micropolitics: The Deliberation Process Within the Commission

> *Thesis: The decision-making process within the ethics commission can be understood not so much in terms of a free discursive coordination, but rather as a 'micropolitical' network of interactions. Major explanatory factors are the actors' strategies and coalitions between the members of the commission, and also processes of power and opportunities to influence the process.*

The aim of the members of the ethics commission is to 'get a voice' themselves, to be heard in politics and among the public. They all consider a cacophony as possibly leading to a loss of influence, both for individual positions and for the commission as a whole. Regarding their aim, the strategy for any individual member cannot be to formulate an argumentation that is as consistent and stringent as possible. Rather, it is important to gain influential 'coalition partners' within the commission. In order to find them, one does not have to develop a sharply profiled position on ethics; one has to act strategically. One has to be prepared to find common ground among different positions, or to make do without putting forward one's own arguments if this would provoke a new round of fundamental debates. One needs to reduce oneself to a fictional canon of 'expressables,' to the set of (internationally established) patterns of argument which include all 'acceptable' arguments. The members consciously exclude decidedly individual positions which would not find majority support within the -commission or be suitable to serve as a basis for coalitions.

> There was a long discussion, and then eleven members of the Bioethics Commission agreed on a positive recommendation. Eight members, of which I was one, rejected that position. We had very different reasons for doing so, and I wasn't really able to explain what my reasons were. I have two reasons which aren't specified in the paper itself. Well, you could even say I had a third position, or something like that... My main reason was that I feel very uncomfortable, because of my conception of democracy, if the EU Commission forces member states to take decisions on matters about which individual states have not had a proper debate – and when the Commission itself is well aware of this fact. Austria has not really discussed the question of research on human embryonic stem cells. (...) In the UK a debate has been in progress for the last twenty years, they have the necessary laws and regulations, and the British can justify a decision to conduct research on human embryonic stem cells because they have thought it through and discussed it adequately. If we decide to do the same just because we don't want to be left behind economically, I have serious ethical reservations. I would say: what are we going to sacrifice next? (Commission Member II).[8]

During the deliberation process, a 'sub-politicisation' takes place in the shape of a loss of importance of 'official' political arenas to the benefit of smaller areas of deliberation that were not decided on beforehand. The discussion keeps moving out of the 'big' commission into meetings of working groups made up of 'likeminded people,' and ultimately to talks and attempts to find agreement in a fully informal setting (such as a private talk in a restaurant). The interviewees described the formulation of a group position as a prolonged process involving tactical moves, strategic considerations, the reaching of multiple compromises etc.

> How many evenings do you think we spent, sitting around trying to find the right way of putting it to make sure everyone was happy? Thinking, for example, we can't say that because some of the esteemed colleagues might take it the wrong way, or...you know what I mean? We sat there for hours, whole evenings, weeks. I'm not exaggerating. We invested an enormous amount of time (Commission Member X).

The supporting texts for position A and position B show a strategic element. B can easily be understood as a compromise between a religious fundamentalism and the kind of scepticism highlighting the inconsistency and the temporal limitation of the safety measures supported by A. There is a range from a Catholic moral theologian insisting on the absolute ontological, moral and juridical status of an embryo, to a patho-physiologist criticising, from a natural science point of view, the concentration of research on embryonic stem cells only. However, ontological and metaphysical arguments are placed in the background. Instead, a risk assessment highlighting elements of societal ethics is put forward. Second, position B refers, to some extent, to the 'slippery slope' argument, assuming that any form of liberalisation of stem cell research will lead to the production of embryos for research, and, therefore, will foster an instrumental attitude toward human life. In other words: position B omits a determination of its position by means of fundamental ethics, which would have to go with a considerable sharpening of the argument. This is explicable in strategic terms. Since more members obviously supported position A, it would have been an additional weakening for position B if it had further split up. From the point of view of content, however, such a split would have been logical, since within position B the tendency to *fundamentally* reject the issue is combined with the tendency, from the perspective of research, to *temporarily* reject it.

Excursus: The Pitfalls of Strategic Action – Unintended Results of Actions

The statement of the bioethics commission cannot be understood simply as a strategically balanced compromise. Unintended results of actions impact both the result of the group's internal deliberations and the political decision-making. Therefore, the result of the deliberation in its concrete form cannot be predicted.

The example of the deadline regulation is instructive with regard to unintended results of strategic action that have an impact on the level of political decision-making. In the commission statement, position A argues for restricting the use of stem cells to those lines existing prior to a certain deadline, in order not to stimulate the production of supernumerary embryos by IVF. The German Ethics Council had previously issued a similar regulation. In the Austrian case, this regulation was proposed by a member of the commission who himself, for ethical-practical reasons, did not consider the regulation to be necessary. Rather, the intention was to support position A by means of a partial acceptance of the rejection by more sceptical members of the commission. This tactically motivated proposal of a deadline, however, developed during the further discussion within the commission into an important point of identification for the fraction of sceptics. In the face of this example, the warning of a 'slippery slope' seemed to be substantiated in a paradigmatic way. According to this view, the deadline regulation could represent, so to speak, a way of making research possible after all. Given the assumption that the quality of the stem cell lines developed prior to the deadline would turn out to be insufficient, one would not be able to avoid slippage up to the complete liberalisation of embryo research (and, hence, for the liberalisation of therapeutic cloning). There would be no place any more for ex-post fundamental ethical concerns. Hence, the deadline regulation served – against the underlying intention – not only as a catalyst to crystallise the sub-group of critics. In the end, it also developed into a central lever for politicians wishing to reject stem cell research by highlighting the lack of safety guarantees.

A 'Culture of Productive Disagreement' Rather Than a 'Battle of Cultures': Dealing With Disagreement Within the Commission

> *Thesis: The enforcement of ethical positions which are not revised fundamentally in the process of bargaining continues to be a subject of strategic action. The anticipated impossibility of reaching agreement has to rely on a professional 'culture of productive disagreement.'*

Our previous remarks should have made clear that we do not interpret the commission as a theatre stage on which a spectacle of bioethics takes place according to a previously well-known script. Even if the recommendation of the commission does not represent the result of a free, unconstrained discursive interchange dominated by the better argument but rather a result of strategically oriented action, of coalitions, and of unintended consequences, it should be clear that the concrete result is not predetermined. In the context of our discussion about the function of dissent, we are more interested in the question of whether the assumption of an overall consensus within the expert group could be realistic. However, this question can definitely be answered only negatively. All members anticipated that they would be unable to agree on the ethically challenging issues.

> Well, I don't think you can complain about the fact that everyone has his own point of view. After all, the commission was put together in such a way as to ensure that a wide variety of viewpoints are represented. (...) Discussions are all very well for young people up to the age of 24, but after that you have to stop discussing and make up your mind. Don't misunderstand me: I don't think there is any difference in this respect between our commission and any of the others that get set up to examine other issues (Commission Member IX).

Not only is it anticipated that the members of the group will disagree with each other, they are actually expected to do so. Disagreement seems to be a kind of guarantee of the quality of expert discussions. It is, firstly, an inevitable consequence of the Chancellor's appointments policy, which has to be seen to be credible, and, secondly, an inevitable consequence of the specific professional socialization and individuation of the experts, who are usually university professors with the corresponding *habitus*. Every national debate between experts is also embedded in an international context. The orientation towards influential discourses, which in the end means bringing into play established models of argument and existing basic positions, is very important in the production of an output with a certain content. This awareness of the discursive pre-structuring of debates in expert bodies also involves being aware that the important thing for the concrete result is less a matter of weighing up theoretical considerations and more a question of constructing majorities in a strategic-political way. Therefore, the degrees of flexibility present in an expert discourse relate less to the cognitive-normative level than to the level of micropolitical action.

> I didn't see any evidence to suggest that, let's say, any significantly new arguments were developed within the group, or even that they could have been developed. By that I mean any argument that had not already been gone through in the international bioethics debate over a long period. The only real question was that of the balance of forces in the commission, of how the different groups would line up (Commission Member V).

However, the fact that debates in the commission are conducted in accordance with arguments which are established and already worked out does not mean that the positions of the different groups are predominantly characterized by logical stringency and normative consistency. The Bioethics Commission's position is not just a local reflection of the range of ethical theories available and of views held by different parties on the national and international level. In the final analysis, the 'mix' of individual positions that emerges at the end depends on the concrete processes of group-formation. Normative consistency is far from being the primary structuring principle. Rather, the opposing positions seem to be derived from a diffuse but basic position which, in the course of the negotiations, is enriched by other arguments that have the capacity to command majority support.

> Well, if you look at where it originates, what happens is simply that the arguments attach themselves to these basic positions like burrs, they just fit. And then those who are in favour put forward one view ... they just try to strengthen their position with a canon of arguments. And the other members of the commission try to make other arguments sound more convincing, and in the end what you've got on the table are a couple of positions that are more or less supposed to reinforce each other – or not, as the case may be. The whole thing grows, like strudel pastry (Commission Member IV).

BIOETHICAL CONTROVERSIES AND POLICY ADVICE

The desire to be politically influential or to provide a forum for public debates needs, even if not real consent, at least a clear setting out of the main positions. This implies that the members feel a certain pressure on themselves to reach an agreement. Only then, as we have already argued, does the necessity arise for 'micropolitical' interactions like reaching a compromise, building coalitions, maneuvring. Strengthening one's own ethical position remains the highest aim among the commission members; there is no interest in maximizing one's influence at any price, for instance by moving to the majority fraction. On the contrary: a (strategically motivated) abandonment of ethical positions will take place only if the fundament of the respective ethical orientations remains unchanged – *in the result* (for example, when the Catholic theologian joins a position whose opposition to stem cell research is not determined by theological-ontological arguments but founded in a pragmatic way).

The interaction about divergent positions within the commission takes place in the context of a 'consensual dissent culture.' The commission practice even leads to a defusing rather than to sharp profiling of the different positions. The interviewees said they explicitly wanted to avoid a 'battle of cultures' within the commission's work. The condition required for this productive disagreement is a generalised willingness to cooperate, a kind of professional friendliness. These are elements of a widely accepted academic identity and help individuals to reflect on the contingency of their own position.

> This is one of our exercises: to explain the premises and background assumptions of your own viewpoint in such a way that the person you are trying to convince can at least empathize, and doesn't experience the exchange as a battle of cultures (Kulturkampf). That's not such an easy thing to do, even among the members of the commission. You have to be objective and friendly at the same time, because these people all know each other. Some of them have close professional ties and regular contact with one another. And it's also an interesting learning process, taking a different view on a question you think is very important in a commission, and then the next day in another context continuing to work on a perfectly friendly basis with the same people. You have to practise quite hard to get that right (Commission Member IV).

In this passage, one can see that the orientation towards 'productive disagreement' is largely explained in terms of the close internal connections within the circle of experts, and so in the final analysis in terms of the fact that the world of experts (especially in Austria) is small and that the people who make up this world are dependent on one another. In the following extract from an interview, the explanation offered has more to do with professional socialisation: in the course of time, every scholar is forced to recognize the contingency of positions from which scientific observations are made and of scientific knowledge.

> And I think, yes, you just learn that there are different ways of looking at things, and depending on where you stand you will see things quite differently. Don't you agree? I think you also have to learn to deal with – I'll say it again – constructive disagreement. And in this respect, I must say, of course ... or rather, the idea wasn't completely unfamiliar to me. I think you learn that there isn't always just one opinion, or just one truth (Commission Member VIII).

One thing which obviously belongs to this expert culture is the critical deliberation and balanced reflection rather than the sharp profiling of one's own ethical position. Last but not least, this leads to the impression that both positions look very similar in relation to the structure and mode of argumentation – even if contradictory recommendations are given as a result.

The cooperative manner of working within the commission is widely illustrated by the fact that members were partly willing to work with representatives of the 'antagonistic' position in order to enhance formulations and arguments. It is most important for the experts, we would argue, to secure the professional authority and the persuasiveness of the whole commission in the eyes of both politics and the public. Short-term political influence is only a secondary objective. (A badly organised counter-position would have been useful for one's own fraction in the public debate at first glance, but this consideration implies that the political enforcement of one's own position soon reaches its limits).

In the long run this way of acting seems to be absolutely functional from the perspective of the expert commission. Ensuring the scientific authority of the commission is in general an important prerequisite of remaining influential in the field of policy. If the commission had disqualified itself in the eyes of the government and the public by a lack of accuracy or implausible arguments, this would have reduced the opportunities for influence of the individual positions represented. This kind of risk exists not only for individual members but also for the commission as a whole. These considerations lead us to the conclusion that disagreement both calls for and enhances the capacity to cooperate. Due to reasons of maintaining the authority of science, it is necessary not to blame the 'ethical enemy.'

"Ethics is Dynamic": The Relationship Between Knowledge and Values

> *Thesis: Bioethical recommendations are bound to refer to the context of specific scientific facts. This technical knowledge, however, requires interpretation before it can become significant for ethical consideration. For this reason it makes no sense to differentiate strictly between (objective) knowledge and (subjective) values.*

Profound knowledge of current biological and medical discussions is considered a basic precondition for a well-founded recommendation by all members of the commission. How to assess the therapeutic potential of human embryonic stem cells? How to assess their potential to develop and differentiate themselves? Is it necessary – from a scientific point of view – to judge between the protection of the embryo and research freedom, in other words, are the therapeutic hopes connected with stem cell research realistic at all?

On this account, the significance of the current research findings becomes a subject of interpretation within the commission, in the process of ethical deliberation and bargaining between positions. An important question at the beginning of the discussion which is relevant for all members, natural scientists as well as social scientists, might be: does the latest article in *Nature*, in which analogies from experiments with mice are drawn to human embryonic stem cells, promise too much? In an attempt to base the ethical decision on a solid and 'value-free' foundation in this way, the publication policies of the influential peer-reviewed journals or the capability of interpretation of the participating natural scientists may influence the ethical judgement to a certain extent. From a more formal perspective, one can certainly characterise this as a division of labour between the different disciplines within the commission. First, at the beginning of a new topic, human geneticists and medical doctors from various fields give a short overview of the state of international research. Next, the lawyers inform their colleagues about the juridical regulations on the national as well as the European level. However, it would be wrong to assume that this kind of division of labour could be assigned to the bargaining processes – and, hence, to presume unreflectedly a differentiation between knowledge and values. Such a fallacy is obvious in the case of bioethics commissions because the phase of ethical judgement seems to follow after a phase where technical knowledge is arbitrated by the natural scientists. However, after careful consideration it is quite evident that the facts are not given first, after which the assessment follows; nor is it the case that scientific facts are evaluated according to individual preferences. In reality, knowledge and values are indistinguishable in this process. What the individual members or the whole expert group accept to be valid scientific or technical knowledge (in other words: real facts) is in fact subject to interpretation. In reality, the social *as well as the natural scientists* are unable to decide whether the articles published in renowned peer-reviewed journals are fully in accord with the relevant quality factors. A deeper investigation about their scientific quality would take far too long. Even if such an investigation were possible: in the current phase of research, analogies and extrapolations are necessary to provide a careful assessment of the feasibility and the potential benefits of stem cell research. And even if it were possible to state these issues more precisely, one has to remember that current knowledge can become outdated very fast. In other words: the scientific facts are put visibly into perspective in multiple ways. What is accepted as a real fact depends on the factual knowledge, the work, and the form of presentation of the participating scientists, and so it depends on the individual assessment of the facts presented, a process which is essentially interlinked with personal dispositions. These dispositions are defined as a bundle of imaginations and expectations that governs the way of perceiving and interacting. In this way dispositions are interlinked with societal discourses, 'ideologies' and so on, and they in turn influence and shape these structures. In the concrete case, the idea of dispositions means that the acceptance of something as a fact also means a judgement about the credibility of the experts who provide the information, or a view about the relevance of a specific scientific and academic tradition which is well-defined against others, and so on. Factual knowledge, from this perspective, becomes a product of complex social bargaining processes. This does not imply that in ethical discourse all knowledge is relative or that values dominate knowledge. Of course, an ethical assessment of biomedical practices cannot work without medical and human genetic knowledge, which defines a specific problem as relevant to a decision. However, the dependence of ethics on science goes further. According to the statements of different experts, ethical expertise is dynamic: ethical considerations depend on the current state of scientific research in the field of genetics and biomedicine.

> Ethics has something dynamic about it, it's a process in which, because knowledge is always advancing (and that's the basic problem, this growth of knowledge), one's view of the consequences can change because there are other consequences, or the consequen-ces can be better controlled, or whatever it may be. So, it is unbelievably dynamic. And I

think that this aspect, and the fact that it's open, should be very strongly emphasized – in a preliminary statement of some sort. And then if there are some objective changes, you have to do more work on the issue (Commission Member VII).

This element, the idea of knowledge as something incomplete and provisionally acknowledged, contributes to an awareness that the question being addressed is in principle open and temporary. At the social level, this awareness of the provisional nature of all expertise tends to defuse conflicts. As a result, the culture of expertise described above as 'productive disagreement' becomes possible.

So much is in flux, quite simply as far as the state of research, the state of empirical research, is concerned, that you can't easily adopt any kind of fixed … or rather, that there's absolutely no need to defend any kind of fixed position whatever the cost (Commission Member X).

That natural scientific knowledge is dependent on interpretation makes it plausible that the attachment of the commission members to the two positions cannot be traced back to disciplinary boundaries or to the distinction between the natural sciences and the humanities (for example: scientific belief in progress versus theological scepticism). Position A was in large measure developed by a theologian on the basis of ethical deliberations. It was supported by human geneticists, less so by medical doctors. Position B was supported by the Catholic moral theologian, lawyers and philosophers, the haematologists and the pathologist – almost the whole spectrum of the disciplines represented in the commission. The sceptical position of the medical doctors, for example, was influenced by the unredeemed promises of the benefits of biomedicine (for example, of somatic gene therapy) and by their own experiences with (animal or human) adult stem cells. Thus especially the medical doctors who have specialist research experience and have an unchallenged high reputation in the commission took a particularly critical position.

At the same time, the thesis of a close link between knowledge and values gives us a possible answer to another question. It is often asked whether bioethics commissions are a form of scientific policy advice at all. Usually this questioning of bioethics is countered with reference to the scientific status of ethics itself. We do not want to deny this, but from the standpoint of the sociology of knowledge such a question is itself questionable. The commission members move on winding paths through claims and constructions of knowledge which cannot be separated from preferences and values and their modifications, and after a prolonged phase of debate, argument and negotiation they come to different positions – what could be more scientific?

CONCLUSIONS: DISSENT, EXPERT OPINION, AND POLITICAL DECISION-MAKING

After the Austrian expert commission had published its divergent recommendation on the case of stem cell research in May 2002, the minister responsible, Elisabeth Gehrer, went to Brussels and voted against the funding of research in the field of stem cell research. This means that she followed the recommendation of the explicitly research-critical minority (Position B). But her statement also referred to the arguments of Position A. The minister argued that so far there were no guarantees on a European level that the deadline regulation voted for in Position A could serve as a morally acceptable form of stem cell research.

Not surprisingly, the commission members have taken very different views of this political decision. While those who were in agreement with Position B feel validated in their ethical judgment and welcome the decision, some representatives of Position A suspect that the political decision was influenced by different stakeholders (first of all the Catholic church) rather than by reasonable and objective considerations. However, all experts agree that the political function of the bioethics commission is (and should be) primarily a consultative one. No member of the commission considers the ethical recommendation as an authoritative instruction for the government to decide in a specific way. In the light of fundamental expert disagreement, traditional ideas of an advanced expert rationality or purely rational expert judgements that create inherent necessities for politics are not convincing any more. The bioethical experts, the politicians and, last but not least, the public know this to be the case – and it can be seen in the inability to agree within the expert commissions.

The divergent recommendations of the bioethics commission did not lead to a lack of legitimisation for the political decision. Against the background of the result of the ballot, which produced only a narrow majority (and this has a meaning only on a symbolic level), both fundamental options were open to the politicians. Moreover, the rejection of funding stem cell research could be plausibly substantiated with reference to both the pro and the contra position. In any case, the ongoing necessity for a political decision came to the fore. The scientific advice did not reduce the political autonomy of the decision. Politics could refer both to the expert knowledge on which the decision was based, and avoid the suspicion that it had acted without expert authority or decided in an irrational manner. The expert dissent did not result in discussions about new forms of political decision-making – for example by participatory instruments of policy advice.

Nor does expert disagreement pose a threat to the commission members themselves. Even though it is rational from an individual expert's point of view to try to maximise one's own 'ethical faction' by acting strategically, in case of doubt experts prefer to ensure the 'legitimatory surplus' of science, the legitimacy of their professional authority, and their expert status rather than to maximise short-term political influence. Even though the ethical interaction processes display the characteristics of a 'micropolitical' network under the constraints of coalition building, the appointment of the bioethics commission in an era of permanent expert dissent does not mean that the boundaries between scientific expertise and decision-making become blurred (see Latour 1999; and, with a focus on the relationship between expertise and the public, Nowotny et al. 2001). Rather, the example of the bioethics commission shows that the "scientification of politics" does not mean a depolitisation of decisions and factual issues, or a rationalisation in the broader sense of manufacturing a lack of ambiguity by insisting on consent and subsequent imperatives to decide (Weingart 2001b). Especially in bioethical debates, expert disagreement is obvious and permanent. It would be wrong to understand expert dissent as politically dysfunctional. On the other hand, "politicisation of science" does not mean a complete replacement of discourse by strategy or the complete metamorphosis of scientists into political actors. The limits of this politicisation of science are set by professional and scientific policy considerations.

It is another question whether it is helpful to analyze the current boom of scientific policy advice from a perspective proposed by systems theory, which draws the boundaries between science and politics according to unambiguous codings related to specific systems (see Luhmann 1992; Willke 2003). However, our example shows that, on the micro-level of structuring individual actions and interactions, the divergence between politics and science cannot be understood in terms of substantially different types of rationality, for example between forms of action seeking agreement and truth, on the one hand, and powerful strategic action on the other. Micropolitical theories of strategic organisation analysis teach us that every kind of action in the context of organisations, institutions, networks etc. involves coalition building, compromise formation, processes of negotiation and strategic action. However, strategic organisation analysis as well as micropolitical approaches also need to ask how actions are embedded within specific contexts, whether shaped by 'system-specific' or professional frames. These systemic frames come to the fore in divergent aims governing these actions. While it is the 'rationale' of the political system to maximise power by increasing the capacity to act and to put forward reasons in order to justify political decisions vis-a-vis the public, science concentrates on protecting its professional authority and claiming to provide implementable expert knowledge.

* *Institute of Technology Assessment of the Austrian Academy of Sciences, Vienna, Austria*
** *Institut für Sozialforschung an der Johann Wolfgang Goethe-Universität, Frankfurt am Main, Germany*

NOTES

[1] For more details, see Bogner and Menz 2002.

[2] For the differences between the concept of micropolitics and the strategic organisational analysis, see Friedberg 2003.

[3] Political commentaries in different Austrian newspapers criticised the fact that only experts were appointed to the commission, see Gmeiner and Körtner 2002: 167ff.

[4] The title of this conference was *Genetische Daten: woher, wohin, wozu? (Genetic Data: Where from? Whither? What for?).* The organiser was a public relations agency, and the conference was financed by the Austrian Council for Research and Technology Development.

[5] This was the first citizens' conference concerning a bioethical topic. A consensus conference on the topic of climate change was carried out in Austria in 1997.

[6] Our ongoing project entitled *Bioethical decision-making and political legitimation. Citizen's participation and Bioethics commissions as instruments of policy advice in Germany and Austria* addresses similar questions. This project is financed by the German Federal Ministry of Education and Research and started in May 2004.

[7] The full text is available athttp://www.bka.gv.at/bka/bioethik/englisch/index_empfehlungen_engl.html.

[8] The following quotations derive from interviews with the members of the Austrian Bioethics Commission conducted by Alexander Bogner and Erich Griessler (Institute for Advanced Studies, Vienna) in autumn 2002.

REFERENCES

Beck, U. (1993), *Die Erfindung des Politischen. Zu einer Theorie reflexiver Modernisierung*, Frankfurt a.m.: Suhrkamp.

Beck, U. and W. Bonß (eds.), (2001), *Die Modernisierung der Moderne*, Frankfurt a.m.: Suhrkamp.

Bogner, A. and W. Menz (2002), 'Wissenschaftliche Politikberatung? Der Dissens der Experten und die Autorität der Politik', *Leviathan* **30**: 384–99.

Bonß, W. (1995), *Vom Risiko. Unsicherheit und Ungewissheit in der Moderne*, Hamburg: Hamburger Edition.

Bundeskanzleramt (ed.), (2001), 'Verordnung des Bundeskanzlers über die Einsetzung einer Bioethik-kommission', *Bundesgesetzblatt für die Republik Österreich* vom 29.06.2001, pp. 1283–4.

Crozier, M. and E. Friedberg (1977), *L'acteur et le système: Les contraintes de l'action collective*, Paris: Seuil.

Friedberg, E. (1993), *Le pouvoir et la règle*, Paris: Seuil.

Friedberg, E. (2003), 'Mikropolitik und organisationelles Lernen', in H. Brentel, H. Klemisch and H. Rohn (eds.), *Lernendes Unternehmen. Konzepte und Instrumente für eine zukunftsfähige Unternehmens- und Organisationsentwicklung*, Wiesbaden: Westdeutscher Verlag, pp. 97–108.

Gmeiner, R. and U. Körtner (2002), 'Die Bioethikkommission beim Bundeskanzleramt – Aufgaben, Arbeitsweise, Bedeutung', *Recht der Medizin* **9**: 164–73.

Gottweis, H. (2001), 'Die unsichtbare Kommission', *Der Standard*, 13.12.2001.

Habermas, J. (1971), 'Vorbereitende Bemerkungen zu einer Theorie der kommunikativen Kompetenz', in J. Habermas and N. Luhmann (eds.), *Theorie der Gesellschaft oder Sozialtechnologie*. Frankfurt a.M.: Suhrkamp, pp. 101–41.

Japp, K.P. (1992), 'Selbstverstärkungseffekte riskanter Entscheidungen. Zum Verhältnis von Rationalität und Risiko', *Zeitschrift für Soziologie* **21**: 31–48.

Japp, K.P. (1996), *Soziologische Risikoforschung – Funktionale Differenzierung, Politisierung und Reflexion*, Weinheim and München: Juventa.

Joss, S. and J. Durant (eds.), (1995), *Public participation in science: the role of consensus conferences in Europe*, London: Science Museum Press.

Joss, S. and H. Torgersen (2002), 'Implementing participatory technology assessment – From import to national innovation', in S. Joss and S. Bellucci (eds.), *Participatory Technology Assessment. European perspectives*, London: Centre for the Study of Democracy, pp. 157–87

Kuhlmann, A. (2002), 'Kommissionsethik. Zur neuen Institutionalisierung der Moral', *Merkur* **56**: 26–37.

Küpper, W. and A. Felsch (1999), *Organisation, Macht und Ökonomie – Mikropolitik und die Konstitution organisationaler Handlungssysteme*, Wiesbaden: Westdeutscher Verlag.

Küpper, W. and G. Ortmann (eds), (1992), *Mikropolitik: Rationalität, Macht und Spiele in Organisationen*, Opladen: Westdeutscher Verlag.

Latour, B. (1999), *Pandora's Hope: Essays on the Reality of Science Studies*, Cambridge and London: Harvard University Press.

Luhmann, N. (1991), *Soziologie des Risikos*, Berlin and New York: de Gruyter.

Luhmann, N. (1992), *Die Wissenschaft der Gesellschaft*, Frankfurt a.M.: Suhrkamp.

Nowotny, H., P. Scott, and M. Gibbons (2001), *Re-Thinking Science. Knowledge and the Public in an Age of Uncertainty*. Cambridge: Polity Press.

Nullmeier, F., T. Pritzlaff and A. Wiesner (2003), *Mikro-Policy-Analyse. Ethnographische Politikforschung am Beispiel der Hochschulpolitik*, Frankfurt a.M. and New York: Campus.

Ortmann, G., A. Windeler, A. Becker and H.J. Schulz (1990), *Computer und Macht in Organisationen – Mikropolitische Analysen*, Opladen: Westdeutscher Verlag.

Schelsky, H. (1965), 'Der Mensch in der wissenschaftlichen Zivilisation', in H. Schelsky (ed.), *Auf der Suche nach der Wirklichkeit. Gesammelte Aufsätze*, Düsseldorf and Köln: Diedrichs-Verlag, pp. 439–80.

Schüssel, W. (2001), 'Wir brauchen ein Frühwarnsystem für neue wissenschaftliche Entwicklungen', *Die Presse*, 16.03.2001.

Tacke, V. (2000), 'Das Risiko der Unsicherheitsabsorption. Ein Vergleich konstruktivistischer Beobachtungsweisen des BSE-Risikos', *Zeitschrift für Soziologie* **29**: 83–102.

Weber, M. (1958), 'Parlament und Regierung im neugeordneten Deutschland. Zur politischen Kritik des Beamtentums und Parteiwesens', in M. Weber (ed.), *Gesammelte Politische Schriften*, Tübingen: Mohr, pp. 294–431.

Weingart, P. (2001a), 'Paradoxes of scientific advice to politics', in *OECD Proceedings: Social Sciences for Knowledge and Decision Making*, pp. 79–94.

Weingart, P. (2001b), *Die Stunde der Wahrheit? Zum Verhältnis der Wissenschaft zu Politik, Wirtschaft und Medien in der Wissensgesellschaft*, Weilerswirst: Velbrück Wissenschaft.

Willke, H. (2003), *Heterotopia. Studien zur Krisis der Ordnung moderner Gesellschaften*, Frankfurt a.M.: Suhrkamp.

Winner, L., A. Feenberg and T.H. Nielsen (eds.), (1997), *Technology and Democracy: Technology in the Public Sphere*, Oslo: Center for Technology and Culture.

CHAPTER 3

HARALD HEINRICHS

ADVISORY SYSTEMS IN PLURALISTIC KNOWLEDGE SOCIETIES: A CRITERIA-BASED TYPOLOGY TO ASSESS AND OPTIMIZE ENVIRONMENTAL POLICY ADVICE[1]

INTRODUCTION

Advising those in power is an old business: depending on the historical-cultural context, wise men, gurus, holy men with magic forces, fortune-tellers or (self-appointed) prophets stand as advisors by the side of rulers and political leaders. In modern societies, scientific and technical know-how is ascribed particular rationality, so that politicians today like to surround themselves with scientifically trained experts as advisors, even if they may continue to listen to their private gurus behind closed doors. They hope to obtain instrumental factual knowledge and ensure legitimacy for their decisions in democratic communities. In fact, only good decisions that solve collective problems and are beneficial to the majority of the population find public acceptance and secure the retention of power for those who govern in democracies. And this is the point in politics (Luhmann 2000).

Policy advice has been enormously expanded and differentiated since the middle of the 20th century. There is science-based advice on all political levels, in diversified thematic fields and policy areas: governmental and parliamentary advice (ad hoc or institutionalized), advice to political parties, expert activities, personal advisors of politicians, or think tanks, which also provide the interested public with expertise (e.g., Barker and Peters 1993; Murswiek 1994; Gellner 1995; Cassel 2000; Glynn et al. 2001, 2003). This expansion, however, has not automatically led to more unambiguous decisions and higher acceptance by the public. On the contrary: the paradoxical effect of an expertise / counter-expertise inflation is observed, which promotes both the scientification of politics and the politicization of science (Weingart 1988, 2001). The publicly apparent dissent among experts in many science-based decision-making processes has weakened scientific expert authority and its legitimation function for politics. The traditional 'social contract' between science, politics and the public with (apparently) clear role allocations has thus become brittle (G. Bechmann and Hronszky 2003). The science-politics interaction as part of advisory processes must be renegotiated according to these analyses and adapted to changed social conditions. But there are no easy solutions for the difficult relationship between (scientific) expert knowledge, forming of the societal will and political decision-making.

Sabine Maasen and Peter Weingart (eds.), Democratization of Expertise? Exploring Novel Forms of Scientific Advice in Political Decision-Making – Sociology of the Sciences, vol. 24, 41–61.
© Springer Science+Business Media B.V. 2009

Against the background of these assessments I will reconstruct in this article the extent to which science-based policy advice is prepared to meet the current challenges of the emerging field of 'science – politics – the public' (Krevert 1993): What possibilities for optimization can be identified using the example of government- and parliament-related environmental policy advice in Germany and the US?

At first I will briefly outline the concept of 'pluralistic knowledge society' as the current context of policy advice in both countries. Following this, I will address important findings of social-scientific research on forms, functions and processes of policy advice. After the analytical frame has been set, I will present central determining factors of environmental policy advisory systems with the example of a German-American comparative study. The criteria-based typology serves as an orientation tool for the assessment and optimization of advisory structures. Finally, I will present some design options for policy-oriented knowledge communication as a possible approach towards proactively facing the challenges of the knowledge society in this field.

POLITICS AND EXPERTISE IN THE CONTEXT OF PLURALISTIC KNOWLEDGE SOCIETIES

Policy advice does not take place in a vacuum. The historically given differentiation structures and the distribution of power, the collectively shared and subgroup-specific value beliefs and basic orientations, the respective conflicting interests and knowledge claims are of central significance for the function and process of advice.

It may thus be meaningful to have a panel of wise experts serving politics as remote prompters aloof from the public debate in hierarchically organized societies: supposedly unambiguous knowledge flows from science to politics, which will then take and enforce indubitable – because factually uncontested – decisions. A modern democracy makes different demands on advisory processes: apart from political control, modern societies rely on societal self-control, civil society and the personal responsibility of their citizens. Moreover, they are characterized by a predominantly positively assessed pluralism of values, interests and knowledge. The concept of a pluralistic knowledge society, which I will expound in the following, covers central macrosociological aspects of this change in Western democracies that has taken place in the past century.

The progressing differentiation in modern societies has been described by numerous authors since the end of the 19th century (Weber 1976; Luhmann 1984; Durkheim 1999; Parsons 2000). The focus has been on the respective socio-structural and socio-cultural differentiation modes and the corresponding socio-political integration mechanisms. At the end of the 20th century a shift from industrial societies to information societies and knowledge societies was diagnosed (Stehr 1994; Bell 1996). Especially in the US, but also in other countries with democratic systems and market economies, a growing social complexity, as well as a pluralization of knowledge claims, interests and values took place (Bohmann 1996).

The current society formation in countries like Germany and the US can be conceptualized as a pluralistic knowledge society following these macrosociological analyses (Heinrichs 2002: 4–38). In addition to the growth of social complexity in the socio-structural perspective there is a significant pluralization of values and interests (Inglehardt 1995; Schimank 1996; Sebaldt 1997). Thus, for example, absolute values such as freedom or human rights are differently interpreted depending on the social position and there are a variety of subgroup-specific value orientations. And social interest pluralism manifests itself in that more than 1,500 interest groups are registered in the lobby list at the German Federal Parliament and approx. 18,000 at the US Congress (Sebaldt 1997: 76; Jäger and Welz 1998: 299).

Of special interest in our context is the pluralism of knowledge and science diagnosed by various authors (e.g., Gibbons et al. 1994; Stehr 1994, 2003; Nowotny 1999; Nowotny et al. 2001). Explicit (scientific) knowledge becomes an increasingly important characteristic of pluralistic knowledge societies. More and more sectors of society are based on systematic knowledge, but at the same time the uncertainty and contingency of pluralistic stocks of knowledge is growing. With regard to scientific knowledge, disciplinary differentiation and segmentation must be taken into account in the same way as recent forms of inter- and transdisciplinary knowledge production, which are designated mode-2 and separated from traditional science (Gibbons et al. 1994). Moreover, different forms of knowledge such as professional practical knowledge or cultural everyday knowledge are also considered to be relevant for social design and decision-making processes (e.g., Krimsky 1984; Wynne 1991; Stehr 1994). The heterogeneity of social groups and actors leads to diversified interpretations of reality in a socio-cultural respect, which represent a challenge for 'socially robust' knowledge and decision-making processes (Nowotny et al. 2001).

These reflexive analyses of scientific knowledge and also of policy-advising expertise have thrown light on the conditionality and limitedness of scientific knowledge: its social construction and demarcation, its relativity, its co-produced non-knowledge, its non-determination and uncertainty, as well as its politicization and industrialization and the – unavoidable – influence of basic orientations, value concepts and interests in trans-scientific expert work.[2] The demystification of scientific knowledge claims as well as public expert controversies have changed the social role of (scientific) experts as disseminators of scientific knowledge in practical contexts (Kleimann 1996: 183–215). The fiction of an unrestricted position of 'freely hovering intelligence' (Mannheim 1995) and of an almost inviolable expert status, as propagated in technocratically conceived 'science societies,' is no longer valid in 'knowledge societies' (Kreibich 1986; Stehr 1994; Stehr 2003).

For such a differentiated and pluralized society, the communicative and responsive understanding of state and democracy focuses above all on the integration of so far insufficiently integrated circles of society. Citizen involvement procedures and other participatory instruments have been developed and applied (Zilleßen 1993; Joss and Durant 1995; Renn et al. 1995). In this sense, advisory systems can be understood as part of the functional intersystem networks, i.e., as part of the social integration mechanisms.

Beyond culture-specific differences, the preceding discussion suggests that there are socio-structural, socio-cultural and socio-political master trends on the macrosociological level, which have triggered similar processes of social change in Western democratic societies. This does not mean that there is a complete universalization of social life in countries like Germany and the US. How these master trends are dealt with, for instance, in regulatory processes, remains dependent on national contexts (for the time being). In this regard, within a large-scale comparative study on chemical regulation, Brickman et al. (1985) showed the relevance of political, social and cultural differences between the US and European countries. And new comparative studies on the meso- and microsociological level will have to show the extent to which international harmonizations or differentiations in policy strategies and regulatory practices will be brought about in the age of globalization (Halffman 2003). Nevertheless, the processes of changed macrostructures apply to Germany and the US, which are referred to in this article for comparative purposes. Both countries can be described as pluralistic knowledge societies in this respect.

The outlined conditions have consequences for the organization of policy advice. The classic knowledge transfer model of instrumental policy advice, in which apparently unambiguous knowledge flows to politics to evoke more rational political decisions in a hierarchical society, has become problematic. Instead, a higher differentiation level in advisory processes seems appropriate in order to accommodate the wide range of knowledge claims, value orientations and interests by a reflecting, transparent and democratic management of expertise pluralism. In the following, I will therefore reconstruct central conceptual models and empirical findings on (environmental) policy advice.

FINDINGS OF SOCIAL-SCIENTIFIC RESEARCH ON POLICY ADVICE

Science, politics and the public rely on each other in democratic societies: the science system has systematic knowledge at its disposal, and politicians legitimized by elections make decisions for which they must try to win public support. The detailed basic features of the interaction relationships are historically variable. Habermas, for example, distinguished more than 30 years ago three fundamental models of the knowledge-value relation between science and politics: decisionism, technocracy and pragmatism (Habermas 1964). In the decisionistic model, politics defines values and goals, and science should deliver instrumental knowledge to achieve the goals. In technocracy, science becomes the dominant institution, because science is believed to identify the 'one best way.' Pragmatism finally is according to Habermas a middle way, in which science and politics have an interdependent, discursive relationship and values and knowledge can be related effectively to each other.

After Habermas numerous – conceptual and empirical – studies were conducted on forms, functions and processes of interaction (e.g., Weiss 1974; Badura 1976; Bruder 1980; Wingens 1989; Jasanoff 1990; Nowotny 1993; Renn 1995; Weingart 1999; Rich and Oh 2000). These studies showed the diversity of advisory practice and the bandwidth of the science-politics-public relationship. At this point, I will selectively address those studies which are central to the empirical comparative study of environmental policy advice in Germany and the US.

The two-community approach identified ideal-typical characteristics of science and politics: truth vs. power, theory vs. practice, cognition logic vs. action logic, facts vs. values, abstraction vs. concretion, complex language vs. simplifying language, long-term time horizon vs. short-term time horizon, modifiable models vs. non-recurring life circumstances, principle of reproducibility vs. principle of irreproducibility, substantial rationality vs. instrumental rationality (Caplan 1979; Wingens

1989). These fundamental characteristics of the systems of science and politics were later criticized as being too undifferentiated, since neither science nor politics are uniform actors (Mayntz 1994: 17–18; Murswiek 1994: 106).

Accordingly, a culture-specific variability of policy advice has been diagnosed (Renn 1995). In ideal-typical terms, a competition model (US), trusteeship model (Southern Europe), consensus model (Japan) and a corporatism model (Northern Europe) can be distinguished. For the two countries compared in this article this means: the US model (adversarial) is oriented to scientific expert dispute, in which data interpretation is in the foreground, whereas expert judgements going beyond scientific argumentation are less relevant. This model is based on the assumption of the methodological objectivity of scientific knowledge. The Northern Europe model (corporatist), in contrast, brings experts and political representatives together. The procedures are formalized and conflicts of interests and different possibilities of influence are recognized and dealt with. The experts, who are often close to interest groups, not only operate as data interpreters. They are conferred a special expert status which puts them in a position to introduce trans-scientific expert judgements (see also Brickman et al. 1985: 315).

Apart from the culture-specific variability of the forms of interaction, a differentiation of interaction functions has also been diagnosed (Boehmer-Christiansen 1995). Going beyond the legitimation and instrumental advising of policy, the full range of functions includes arbitration, decision delay, problem solving, persuasion and others. According to these analyses, the forms and functions of policy advice seem to be more multifaceted than abstract models of the science-politics relationship suggest.

Finally, studies were conducted which analysed the processes of interaction between science and politics beyond rationalistic 'ideal concepts' of a linear transfer of knowledge from a micro-perspective. These studies emphasized the significance of the situational context of the decision-making process, the cognitive limits of information processing by decision-makers and the special nature of scientific expertise (Hammond et al. 1983: 288f.). Moreover, they pointed out that the organization- and person-dependent stock of knowledge and the tacit knowledge of the decision-makers are an important reason for the utilization of expertise. Dealing with information is thus closely linked to implicit stocks of knowledge, preceding explicit knowledge and structuring information behaviour. Expertise is consequently just one source of information for political decision-makers, and information processing is context-dependent. Information does not determine the policy decision and information is chosen selectively. From this perspective, policy advice is not to be understood as a linear process, but as a "web of communication" (Rich and Oh 2000: 173f.).

Based on these findings, integrative approaches of dialogistic policy and public advice were developed in the late eighties. Especially in Great Britain, Canada and the US, analyses and recommendations for reorganizing scientific expertise in the political decision-making process were presented (CSTA 1999; Halliwell et al. 1999; Smith et al. 1999; EPA/SAB 2000; OXERA 2000). In these studies, some central elements can be identified that are also related to the functional change of experts and expertise under changed socio-political boundary conditions described above:

balanced committees, nature of expertise, scientific uncertainty, review procedures, transparency, openness, participation, integration of local knowledge, dialogue orientation. The driving force for these modifications lies above all in securing the credibility of science-based decision-making processes.

In Germany, too, in the nineties, proposals for changing scientific policy advice were submitted, which tie less into the decisionistic and more into the pragmatic model (Krevert 1993; Renn 1999a; 1999b). This is not surprising against the background of the corporatist advisory model of Northern Europe which relies on expert judgements.

In those concepts, politics plays a moderating role in order to continuously and systematically relate diverging knowledge claims, value concepts and interests to each other. In this way, consensus should be explored and dissent elaborated to enable social integration. Policy advice is seen as an analytical-deliberative process of knowledge compilation and knowledge assessment. Scientific, political-administrative and civil society actors are to be incorporated in the same way as citizens (Renn 1999b: 544). In summary, it may be stated that innovative advisory procedures in many Western countries aim at more efficiently accommodating the pluralism of knowledge, values and interests in socially complex societies and processing it for decision-making.

The social-scientific findings on policy advice described in this section show that a renunciation of traditional science-politics models and naïve rationality concepts is both empirically observed and normatively recommended. In view of these observations, a traditional knowledge transfer model in hierarchically organized (industrial) societies can be contrasted today with a model of advice as communication and negotiation. The central question for a comparative analysis of policy advice in Germany and the US is thus: How far does current environmental policy advice meet the requirements of integrative policy advice in a pluralistic knowledge society?

PERFORMANCE OF CURRENT ENVIRONMENTAL POLICY ADVICE: GERMANY-US COMPARISON

The topic of environmental policy is particularly well suited for a (comparative) analysis of advisory organization and practice due to the enormous complexity of society-environment interdependencies. Whether issues of climate change, biodiversity, land use or hormonally active chemicals are involved, politicians require knowledge on cause-effect relationships and options for action for informed decision-making. Moreover, the dynamics of these comparatively young policy fields reflects the processes of social change described earlier). Thus, a change from command-and-control approaches to more co-operative policy strategies is seen for both Germany and the US (A. Bechmann 1995: 463f.; Lester 1995: 22f.) This development corresponds to the broader trends of political control and societal self-control that have developed to cope with the complexity of pluralistic knowledge societies.

Both in the US and in Germany, diversified advisory activities have become established for governments and parliaments since the institutionalization of environmental policy. The German Federal Ministry for the Environment, Nature Conservation and Nuclear Safety (BMU) and the US Environmental Protection Agency (EPA) use in-house and agency-based expertise as well as external advice. For advisory systems in the US the Federal Advisory Committee Act (FACA) is central, which was passed by Congress in 1972 as part of the trend towards greater openness in government. This Act was intended to contribute towards reorganizing the then unclear advisory activities to ensure a fair participation of external actors and transparent advisory processes. Central requirements are (Long and Beierle 1999: 4):

- establish a written charter that explains the mission of the committee;
- give timely notice of committee meetings in the Federal Register;
- have fair and balanced membership on the committee;
- open committee meetings to the public, whenever possible;
- have the sponsoring agency prepare minutes of committee meetings;
- provide public access to the information used by the committee;
- grant to the federal government the authority to convene and adjourn meetings; and
- terminate within two years unless the committee charter is renewed or otherwise provided for by statute.

Both science-oriented expert advisory committees and more politics-oriented policy level committees must be implemented pursuant to FACA (Long and Beierle 1999: 5f.) Two of the three panels under consideration in this article, the institutionalized Scientific Advisory Board and the ad-hoc Endocrine Disruptor Advisory Committee fall within FACA.

It may be stated that both Germany and the US have a comprehensive advisory system in the area of environmental policy. In the following, by means of the case studies of seven environmental policy advice systems I will show to what extent the advisory forms enable pluralistic scientific and politically value-related claims to be incorporated into decision-making in a way that is democratic, fair and technically efficient.[3]

In Germany, three institutionalized advisory systems established by the executive, or funded institutionally by the executive, and one ad-hoc advisory system in Parliament were investigated. In the US, three advisory systems in the issue area of 'endocrine disruptors' were analysed.[4] Table 1 gives an overview of the advisory panels and their characteristics.

The reconstruction of advisory practice is based on guided interviews with persons from the panels, representatives of relevant ministries, parliamentarians, representatives of industrial and environmental associations and journalists as well as document analyses. The case study analyses of each of the seven advisory systems, which I cannot address here in detail, show above all panel-specific details that make apparent the context-relatedness of advisory processes. The results of the case studies are predominantly in agreement with the social-scientific findings on forms, functions and processes of interaction presented in the above chapter:

- First of all, more decisionistically (e.g., NAS Committee) and pragmatistically organized (e.g., Enquete Commission) advisory systems can be distinguished on an abstract level following the analytical perspective of the Habermas advisory models. Technocratic policy advice in the sense that science predefines the goal

HARALD HEINRICHS

Table 1: Overview of the advisory panels and their characteristics

Advisory Panel	Characteristics
German Advisory Council for the Environment (SRU)	Scientific expert panel of the German Federal Government; national/European environmental policy
German Advisory Council on Global Change (WBGU)	Scientific expert panel of the German Federal Government: international environmental and development policy
German Council for Land Conservation (DRL)	Institutionally funded council composed of scientists and practitioners; regional/national nature/land conservation policy
Enquete Commission 'Protecting Humans and the Environment'	Commission of the German Parliament composed of experts and parliamentarians; national sustainability policy
NAS Committee on Hormonally Active Agents	Scientific expert panel of the US National Academy of Sciences (NAS) on behalf of the US Environmental Protection Agency (EPA); issue-specific knowledge processing
Endocrine Disruptor Screening and Testing Advisory Committee (EDSTAC)	Commission on behalf of EPA composed of scientists and experts from administrative, industrial, environmental and public-health associations; issue-specific programme development
SAB / SAP Subcommittee on Endocrine Disruptor	Commission on behalf of EPA composed of scientists and experts; issue-specific programme evaluation

- and route and politics only follows cannot be found. However, it should also be noted for future studies that science increasingly is the social system that creates political action problems by its discoveries (example: climate change) and inventions (example: biotechnology), for the solution of which it is then indispensable (G. Bechmann and Grunwald 2002: 114; Stehr 2003). Panels such as the German Advisory Council on Global Environmental Change (WBGU), which largely defines and elaborates topics independent of the political client and puts political action problems on the agenda while simultaneously supplying action knowledge, undoubtedly exert an influence on political objectives. New (scientific) knowledge thus at least preforms the areas in which politics must set goals.

- Below this abstract level of analysis the case studies show that the ideal-typical two-community approach in large parts does not apply to advisory practice: blurring the dichotomies, whether cognition logic vs. action logic, facts vs. values, theory vs. practice or substantive rationality vs. instrumental rationality is nothing unusual in concrete advisory processes in the opinion of the large majority of those interviewed.

- Concerning the interaction functions, further functions of policy advice were identified – pools of ideas/background knowledge (see below) – which confirm studies pointing out that there is a wider range of advisory functions besides legitimation and instrumental knowledge.

- The culture-specific design of policy advice (only) applies to a limited extent, according to the empirical analyses. Although corporatist forms in fact dominate in Germany (confidence in expert judgements, balanced membership on the Enquete Commission) and the competition model prevails in the US (importance of methodological objectivity of science and separation of scientific data interpretation and political judgement), the Endocrine Disruptor Screening and Testing Committee (EDSTAC) is an example of the fact that there are also corporatist advisory forms in the US. In this advisory system values, interests and knowledge are deliberately brought together in a corporatist form to enable rational coping with the topic, which is characterized by political controversy and scientific uncertainty.

- Finally, the case studies indicated that the interaction processes are not rational knowledge transfer processes from person A to person B, but that problem definitions, information and stocks of knowledge are selected, recontextualized and transformed in both directions of the advisory relations. It will have to be further examined in micro-sociological studies how the associated processes of (re-)construction and shifts of meaning take place in detail.

In spite of the variety, variability and context-dependence of advisory processes, four central cross-panel dimensions, with which the advisory systems are confronted both in Germany and in the US, can be identified. A criteria-based categorization of environmental policy advisory systems along these dimensions may serve as an orientation tool for interactions among science, politics and the public. In the following, I will outline the dimensions, the associated criteria, which result from the material, and the categorization of the advisory systems.

Dimension 1: Distance from Politics

Advisory processes cannot be considered separately from the respective political conditions that prevail and reflect the given distribution of power in a particular historical situation. A current example is the Bush Administration's (attempted) influence on advisory structures and panels in the US. Some observers have expressed concern about the efforts of the Administration to install 'advice without dissent,' where panels with formerly heterogeneous membership, so-called 'balanced committees' are restructured to become 'administration-friendly' (Michaels et al. 2002). Even if no relationship of dependence or a direct instrumentalization was diagnosed for any of the panels investigated in the present study, the extent of political influence is the central dimension of environmental policy advice systems.

Distance from politics is determined by planning, appointing and accompanying the panels. The criteria to assess this dimension are focused on the following aspects:

- from the large number of conceivable problem areas politics narrows down the topics to be potentially dealt with ('political agenda setting');
- policy-makers determine the panel's possibilities and limits of action by strategic council orientation;
- the stocks of knowledge to be referred to are selected according to technical but also political criteria;
- policy-makers select the disciplines from which they expect to receive the necessary expertise;
- policy-makers appoint the experts according to subject-related criteria but also according to political and in part personal preferences;
- participation of policy-makers in the production of expertise is a far-reaching possibility of exerting political influence.

The advisory systems investigated can be categorized based on these criteria from 'close to politics' through a medium position up to 'distant from politics.'

Panels organized close to politics such as the Enquete Commission in Germany or the Endocrine Screening and Testing Advisory Committee (EDSTAC) in the US are considerably marked by politics. As the following statements of policy-makers and experts interviewed for this study show, the advisory process is politically influenced from the selection of the problem areas to be dealt with up to participation in the preparation of expert reports:

> ... I have always intended for the EDSTAC to address issues that lie at the complex interplay between science and policy. Let there be no doubt that the EDSTAC's recommendations must be firmly and thoroughly grounded in sound science. However, the issues that EDSTAC is charged with addressing also have a policy dimension. The interplay between science and policy is another reason why I believe it is worthwhile pursuing a consensus objective with a group that is as broad and diverse as the EDSTAC ... (policy-maker, EPA).

On the one hand, policy-relevant expertise can be produced with concrete problem solution competence due to the close co-operation between experts and politics. On the other hand, closeness to politics can adversely affect the rationalization and legitimation power of scientific expertise in public discourse:

> ... Politics has to like the output. That is clear. Science has a serving function, that means politics defines the event... (expert, Enquete Commission, translated).

Advisory systems positioned in the middle are also politically influenced by appointment procedures and panel design, but these panels gain strength in rationalization and legitimation since expertise production takes place independent of the political client. These panels – e.g., the German Advisory Council for the Environment

(SRU) – are thus positioned between the poles of decision preparation close to politics and scientific enlightenment:

> ... but then we expect concrete policy recommendations and concrete discussions of policy-making (...), they don't always want to support practical policy-making. Instead, they wish to be more abstract and more general in their recommendations ... (policy-maker, German Environmental Ministry).

Panels distant from politics, finally, are least marked by the political calculations of the clients. Although the institutional boundary conditions are also set politically here, these advisory institutions have considerable scope in organizing the process. Distance from politics, which is beneficial due to the legitimation function of scientific independence, is paid for by disadvantages concerning the usefulness for shaping policy. These problems became especially apparent in the case of the NAS Committee on Hormonally Active Agents: no consensus was reached and therefore it could just be used by policy-makers as background knowledge and data collection, but not as an instrument for making and legitimating decisions.

In view of the trans-scientific nature of expertise and the unavoidability of expert pluralism, transparency with respect to the planning, appointment and accompaniment of a panel as well as expert selection appears useful. In the German advisory systems in general the non-scientific, political influences are not made systematically explicit. This is in line with the more corporatist approach. Although especially in the Enquete Commission a wide range of different opinions are incorporated by the party-driven appointment of experts, expert diversity is not above party politics and not intentionally balanced; it results from party-political calculations (see also Brown et al. in this volume). With regard to the other advisory bodies, political influences with respect to appointment and council orientation are insufficiently transparent.

For the American panels, which are closer to politics than the NAS Committee, a greater sensitivity regarding political distance can be observed. The targeted appointment of balanced committees and greater transparency in expert selection enable – particularly in EDSTAC – expert pluralism to be dealt with productively under changed boundary conditions. The Federal Advisory Committee Act, which regulates this process formalization in a binding manner, plays a special role here.

Dimension 2: Policy Function

How expertise is used and processed depends on the – explicit or implicit – allocation of political functions. In total, four central patterns of using expertise can be identified: decision preparation (instrumental), argumentation aid/reference point (with respect to legitimation/rationalization), increasing the pool of ideas and more general enlightenment. What political function(s) is (are) fulfilled by a panel can be assessed by the following criteria:

- positioning of the panel (more policy- or more science-oriented);
- reference level (concrete problem-specific or abstract-generalizing);
- time perspective of the expertise (short-term, medium-term or long-term).

Thus, for example, a panel that is strongly policy-oriented and deals with concrete problems on a short-term basis, such as the Endocrine Disruptor Screening and Testing Advisory Committee, serves to prepare decisions:

> ... How can we say, we were part of the advice, but then turn our back on it? So we had a higher interest in the outcome ... (policy-maker, EPA).

On the other pole of this dimension there is expertise that is more science-oriented and makes abstract, generalizing statements of long-term significance. This expertise is rather used as background knowledge and influences political perception and interpretation patterns only slowly (e.g., German Council for Land Conservation (DRL) or the NAS Committee on Hormonally Active Agents):

> ... this is more (...) indirect, much more indirect a voice (...). The council can make points in the argumentation, serve as reference point and influence the direction of the discussion ... (expert, DRL).

Advisory systems in a socially complex society with differentiated advisory contexts have to fulfil different political functions. This requires a specific organization of advice in each specific case. The heterogeneity of the German advisory system is compliant in this respect with the requirements of a pluralistic knowledge society. But the individual panels and their expert reports are hardly coordinated. Moreover, especially the SRU or WBGU advisory systems reveal the effort to simultaneously fulfil different political functions. In the worst case this may lead to no function being well fulfilled.

The advisory structure of the American Environmental Protection Agency shown by the case study of endocrine disruptors represents a structured and coordinated advisory organization with clear policy functions. Different panel types were appointed function-specifically to generate background knowledge (NAS Committee), develop consensually strategic recommendations (EDSTAC) and evaluate the EPA programmes based on this expertise for decision preparation. In this way, different advice contexts are complied with. The analysis of the American advisory systems points to the fact that clear positioning with respect to advisory services is beneficial for meeting the requirements of differentiated advice contexts and enabling targeted advice.

Dimension 3: Dealing with Pluralism of Knowledge, Values and Interests

Any policy advice in which scientific expertise is related to political problems goes beyond a purely scientific discourse. It is trans-scientific, because the advisory work is marked by the conditions that specific signals are sent to politics (Weinberg 1972; Rip 1985). Therefore, apart from different stocks of knowledge, the value and interest background of the experts is also of relevance in the advisory discourses. The way in which the pluralism of knowledge, values and interests is dealt with in the advisory process may greatly vary: the spectrum ranges from pluralism being a non-topic through mixed forms up to the proactive handling of these aspects. The panels can be assessed on this dimension according to the following criteria:

- pluralism is made explicit in the advisory work in order to be able to deal with it more deliberately (topic);
- pluralism is represented and accepted but not dealt with systematically (mixed form);
- pluralism remains largely implicit (non-topic).

The pluralism of knowledge, values and interests of modern societies is partially taken up by the German advisory systems. The interdependencies of knowledge, values and interests are reflected by most interviewees, but hardly actively discussed in the advisory processes of SRU, WBGU and DRL. The interviewees rather held the opinion that different value- and interest-related positions must be balanced in the course of advisory discussions:

> ... Of course, there are always differences in opinion. Committee members, who claim what is not proven should not bother us. And there are others, who have the opposite standpoint. Somehow we adjust to each other ... (expert, SRU).

Pluralism is thus represented in the advisory discourses. It is accepted and pragmatically dealt with. In the Enquete Commission, the pluralism of knowledge, values and interests is much more apparent to those involved from the very beginning. A consensus is aimed at by discussions and negotiations. But no systematic treatment of the different levels of discourse takes place here either.

This way of dealing with pluralism, which relies on negotiation rather than on reflection, corresponds to the culture-specific model of corporatism and to the special role of expert judgements in Germany.

The relevance of dealing with these aspects as transparently as possible is shown by the American study. A non-topic approach of fundamental value concepts and interests observed as a trend in the NAS Committee appears especially problematic for issues that are characterized by scientific uncertainty and political controversy.[5] The dominating orientation of the American competition model towards the methodological objectivity of scientific knowledge, and the attempt to maintain a demarcation line between facts and values, seems questionable under conditions of cognitive uncertainty and normative ambivalence:

> ... We were asekd how do we felt about it. There was this (...), he was one person of the Troika, he said he had three children and that he cared, because he want to give them a nice environment ... (and other said) ... like, there is no problem, chemicals have improved the life, I don't remember. The ones from industry said that the people tend to make things out of nothing. And one guy from the industry said it's nice that we are together in this room, because I never would have expected that (...) would shake my hand, so he marked me as an activist ... (expert, NAS Committee).

Especially if pluralistic basic assumptions remain implicit or are ignored, this may lead to grave misunderstandings and aggravate the advisory work, as happened within the NAS Committee. The EDSTAC process, on the contrary, has shown that a disclosure of pluralistic claims is helpful for the rationalization of advisory processes. A proactive discussion of these implicit relations also facilitates a clear differentiation of knowledge, uncertain knowledge and non-knowledge. In this way, blurred areas of value- and interest-oriented opinion as well as theoretically known and empirically verified aspects can be better elucidated.

54 HARALD HEINRICHS

Dimension 4: Communication, Interaction, Inclusion

The fourth dimension of environmental policy advice systems relates to the external relations of the panels, which are characterized by typical communication, interaction and inclusion patterns. A differentiation can be made here between input communication (activities for expertise production) and output communication (activities for expertise dissemination). The advisory systems work in an excluding, including or including-dialogistic manner depending on the degree of interrelations with their advice contexts. The categorization of the panels on the dimension 'input communication' is based on four criteria:

- The first criterion determines whether the selection of topics is cooperative or autonomous. Whereas a cooperative topic selection, in which different political actors may be involved, permits a better consideration of political needs, the autonomous selection of topics allows new scientific topics to be brought closer to politics.
- The second criterion relates to knowledge integration. Is a wide range of different knowledge claims taken into account in expertise production or is narrowly defined, specific knowledge used for a detailed analysis?
- The third criterion shows whether access to the advisory process is transparent for the public or whether expertise production takes place behind closed doors. For the generation of response by political, sub-political and medial actors, transparency seems to be as important as the participation possibilities of relevant actors, for example, by making comments.
- The fourth criterion finally concerns the style of communication. Does information acquisition for expertise production predominantly take place in written form via literature, documents etc. or is it complemented by face-to-face communication such as informal exchange of opinion or formal hearings with relevant actor groups?

In the dissemination of results, which is designated here as output communication, the comparative analysis of the seven advisory systems also revealed cross-panel patterns.

- The most important criterion for this dimension relates to the type of advice; is state-oriented policy advice performed or is policy and public advice aimed at, in which target groups of civil society are included in the advisory system?
- In close relation to this, the second criterion concerns the type of knowledge dissemination. Is the aim to disseminate expertise exclusively to the sponsoring clients or to widely spread the results to sub-political actors and the media?
- The third criterion considers whether media work is active or passive. Professional media work is indispensable for effective public communication in modern media societies.
- The fourth criterion finally concerns the style of communication. The dissemination of results is primarily performed by means of expertise reports, complemented by abstracts, executive summaries, target-group-oriented text editing, press releases, etc. In addition, there are also face-to-face activities such as press

conferences, (informal) journalist meetings, workshops for dissemination to individual target groups, etc.

Regarding the requirements of integrative policy and public advice, the analysis of the German advisory systems has revealed weaknesses. The inclusion of central political and sub-political actors like the media is only insufficiently systematized. The science-oriented advisory systems SRU, WBGU and DRL aim at broad-based knowledge integration and selectively involve different actor groups in the advisory process for both input and output communication. But the advisory work is largely non-public and there are only few opportunities for face-to-face communication. Elements such as public hearings or target-group oriented dissemination workshops and information events only take place sporadically. The media work of the advisory systems also appears worth improving:

> ...Yes, it is true that we do not have the resources in our office, that we can or want to do offensive, perhaps even aggressive public relations. We are primarily a scientific advisory body exclusively for the government and we pay less attention to the public. However, we aim at reaching the public ... (expert, SRU).

Although all the panels perform media work, they largely take insufficient note of the mechanisms of media production. This impairs the connectivity to the political and public agenda.

The American advisory systems offer more integrative policy and public advice. Whereas the NAS Committee tended to work in an exclusive manner, the other two panels (EDSTAC and SAB/SAP) were established under the Federal Advisory Committee Act (FACA). The dialogistic-inclusive input and output communication, which is marked by public access and face-to-face communication, has proven useful for policy relevance in the opinion of those interviewed:

> ... From the beginning this was a transparent process. Every time we had a meeting we had at the end set aside time for public input. So there were some people. So the process was tracked by people from outside the committee. The dates of the meetings were made public on the internet, there was nothing done behind closed doors. Consequently, we had feedback from the public throughout the entire process. So there were no surprises in the end. When the final report came out, everybody in the public who was interested knew what this was saying, because it was discussed right from the beginning ... (expert, SAB/SAP Subcommittee).

Even though integrative policy and public advice is resource-intensive (see Brickman et al. 1985) it seems that in complex science-based issues with scientific uncertainty and political controversy the broad inclusion of heterogeneous participants in the end raises the chance for lasting collective decisions.

POLICY-ORIENTED KNOWLEDGE COMMUNICATION – A POSSIBLE WAY?

Based on the criteria-based typology we can state that the environmental policy advice systems in Germany and the US only in part comply with the changed social conditions and the demands of integrative policy and public advice. The dominating corporatist model in Germany is not sufficiently systematized for the requirements of a pluralistic knowledge society. It is inadequately structured with a view to political functions, expert/expertise utilization, dealing with the pluralism of knowledge, values and interests and the integration of demands and actors. The US model, which is complemented by corporatist elements, is more in compliance with the requirements. However, the strong orientation to the methodological objectivity of scientific knowledge seems to be problematic, especially in trans-scientific advice contexts. In sum, it may be stated that many elements of modern policy advice are realized in the advisory systems investigated. But a further optimization of the individual advisory panels and of the advisory system as a whole appears necessary in order to master the challenges of socially complex societies. In the following, some design options will be proposed.

Design options in the sense of 'advisory advice' can be derived from analyses of pluralistic knowledge society, with regard to policy advice and the criteria-based typology of advisory processes, in order to organize advisory systems in a democratically fair and factually efficient manner. In this context it is important to be aware of the transition from a relatively static industrial society, in which supposedly unambiguous knowledge was politically implemented in hierarchical structures, to a process-oriented knowledge society, in which a comprehensive communication network continually takes up, processes and reflects demands. Policy advice is thus less conceivable as one-sided knowledge transfer than as politically initiated, moderated and structured knowledge communication including values and interests. For this purpose, policy advice in the sense of a 'one-way transfer' of scientific expertise to governmental decision-makers must be changed to dialogistic, policy-oriented knowledge communication. The following design options can be formulated for the organization of the advisory panels with a view to the criteria-based typology.

These design options can contribute towards realizing an organization of advisory systems satisfying the demands of pluralistic knowledge societies such as Germany or the US. Even though the advisory processes finally are embedded in the varying

Table 2: Design fields – Design options

Design field	Design options
Distance from politics	Disclosure of political influences, precise task description, transparency in the selection of experts and stocks of knowledge, appointment of balanced committees.
Political function	Clear definition of the political function, function-specific equipment and organization of the advisory panel.
Knowledge, value, interest pluralism	Systematic reflection of fundamental values and interests in knowledge discourses; disclosure of the limits of knowledge, of uncertain knowledge and non-knowledge.
Communication, interaction, inclusion	Stronger inclusion of relevant actor groups, more input-output communication.

Table 3: Advisory steps – Advisory functions

Advisory steps	Advisory function
Orientation advice	Systematic knowledge preparation for orientation concerning new (or existing) problem fields; scientific expertise central; consideration of professional practical knowledge and cultural everyday knowledge by dialogistic input/output communication.
Strategy advice	Development of problem solution strategies; scientific knowledge, professional practical knowledge, cultural everyday knowledge (topic-dependent); broad-based policy recommendations for decision preparation.
Evaluation advice	Evaluation of programme efficiency and goal reaching; scientific expertise central; professional practical knowledge and cultural everyday knowledge complementary.

political, social and cultural traditions, the further development of the existing advisory systems relates above all to a more sophisticated differentiation, systematization and structuring of previous advisory processes in order to achieve a higher degree of context-sensitivity. To what extent dynamic processes of globalization and transnationalization stimulate an assimilation of advisory procedures at least in Western democracies is an interesting research question for future comparative analysis.

Besides optimizing the individual advisory systems, policy-oriented knowledge communication also aims at a better coordination of the individual advisory processes to avoid duplication of work, overlapping and inefficiencies, and to structure different advisory aspects more clearly. Three steps of policy-oriented knowledge communication can be differentiated, which form a joint communication network and continually take up, process and evaluate scientific and social demands:

The design options for the organization of advisory processes and the three-step concept of policy-oriented knowledge communication provide a contribution to the structural adaptation of the advisory system to the social complexity and pluralism of modern societies. More participation is not advocated as an end in itself, but with the aim of a function-specific integration and coordination of knowledge, values and interests by adequate participation of scientific, political and sub-political actors and citizens. It is hoped that a targeted organization of environmental policy advice enables a higher differentiation level beyond technocratic constraints and post-modern arbitrariness in advisory processes.

Universität Lüneburg, Germany

NOTES

[1] This study is based on my doctoral thesis, which is published under: Heinrichs, H. (2002): *Politik-beratung in der Wissensgesellschaft. Eine Analyse umweltpolitischer Beratungssysteme,* Wiesbaden.

[2] As an overview see: Felt et al. 1995: 114–48, Maasen 1999: 45–50. For specific aspects of the social conditions of science and science for policy see: Weinberg 1972; Nelkin 1979; Knorr-Cetina 1985; Jasanoff 1990; Cozzens and Woodhouse 1995; Gieryn 1995; Martin and Richards 1995;.

[3] Advisory system is defined here as the action and communication relationship of actors directly and indirectly involved the advisory process.

[4] In the USA – as in Europe – more than 87,000 synthetic substances are in use as industrial and agricultural chemicals. More than 1000 compounds are added every year. The question of whether and how specific chemicals adversely affect humans and animals is thus of high relevance. Since the 1960s, experts have pointed out that synthetic substances can have carcinogenic, mutagenic and teratogenic effects. The politicians responded and imposed numerous regulations in order to reduce or completely avoid the application of individual chemicals. However, the political measures related to current toxicity, carcinogenicity, mutagenicity and teratogenicity alone. In the past two decades, however, numerous field studies and laboratory experiments have been carried out, which suggest subtle effects of synthetic substances on the hormonal system of humans and animals. Numerous clinical pictures ranging from reproduction disturbances through neurobiological effects up to impairment of the immune system are related to the so-called 'environmental endocrine hypothesis.' The test procedures so far used do not pay attention to the effects of chemicals potentially disturbing the hormonal system. Due to the enormous complexity of the problem and the continuing scientific uncertainties, it is possible for the actors involved to come to different conclusions concerning risks and necessities for political action. (Colburn et al. 1996; Krimsky 2000).

[5] In this regard see also Hilgartner (2000). He demonstrates in his work on the National Academy of Sciences to what extent the official 'face' of the NAS in the foreground differs from the 'internal' production prosses in the background.

REFERENCES

Badura, B. (eds.), (1976), *Seminar: Angewandte Sozialforschung. Studien über Voraussetzungen und Bedingungen der Produktion,* OXERA *Diffusion und Verwertung sozialwissenschaftlichen Wissens,* Frankfurt a.M.: Suhrkamp.

Barker, A. and B.G. Peters (eds.), (1993), *The Politics of Expert Advice. Creating, Using and Manipulating Scientific Knowledge for Public Policy,* Pittsburgh: University of Pittsburgh Press.

Bechmann, A. (1995), 'Umweltpolitik als gesellschaftlicher Lernprozeß – Erfahrungen aus 25 Jahren Umweltpolitik in Deutschland', in L. Steubing (ed.), *Natur- und Umweltschutz: Ökologische Grundlagen, Methoden, Umsetzung,* Jena: Fischer, pp. 460–80.

Bechmann, G. and A. Grunwald (2002), 'Experimentelle Politik und die Rolle der Wissenschaften in der Umsetzung von Nachhaltigkeit', in K.-W. Brand (ed.), *Politik der Nachhaltigkeit,* Berlin: edition sigma, pp. 113–29.

Bechmann, G. and I. Hronszky (eds.), (2003), *Expertise and Its Interfaces. The Tense Relationsship of Science and Politics,* Berlin: edition sigma.

Bell, D. (1996), *Die nachindustrielle Gesellschaft,* Frankfurt/M.: Campus.

Boehmer-Christiansen, S. (1995), 'Reflections on scientific advice and EC transboundary pollution policy', *Science and Public Policy* **22**, 3: 195–204.

Bohmann, J. (1996), *Public Deliberation. Pluralism, Complexity, and Democracy,* Cambridge, MA: MIT Press.

Brickman, R., S. Jasanoff and T. Ilgen (1985), *Controlling Chemicals: The Politics of Regulation in Europe and the United States,* Ithaca, NY: Cornell University Press.

Brown, M.B., J. Lentsch and P. Weingart (2006), 'Representation, expertise, and the German Parliament: A comparison of three advisory institutions', in S. Maasen and P. Weingart (eds.), *Democratization of Expertise? Exploring Novel Forms of Scientific Advice in Political Decision-Making – Sociology of the Sciences, vol. 24,* Dordrecht: Springer, pp. 81–100.

Bruder, W. (1980), *Sozialwissenschaften und Politikberatung,* Opladen: Westdeutscher Verlag.

Caplan, N. (1979), 'The two-communities theory and knowledge utilization', *American Behavioral Scientist* **22**: 459–70.

Cassel, S. (2000), *Politikberatung und Politikerberatung. Eine institutionenökonomische Analyse der wissenschaftlichen Beratung der Wirtschaftspolitik,* Bern: Verlag Paul Haupt.

Colburn, T., D. Dumanoski and J.P. Meyers (1996), *Our Stolen Future,* New York: Dutton Signet.

Cozzens, S.E. and E.J. Woodhouse (1995), 'Science, government, and the politics of knowledge', in S. Jasanoff, G. Markle, J. Petersen and T. Pinch (eds.), *Handbook of Science and Technology Studies,* London: Sage Publications, pp. 533–53.

CSTA – Council of Science and Technology Advisors (1999), *Science Advice for Government Effectiveness (SAGE),* Canada.

Durkheim, E. (1999), *Über soziale Arbeitsteilung. Studie über die Organisation höherer Gesellschaften,* Frankfurt a.M.: Suhrkamp.

Environmental Protection Agency (EPA)/Scientific Advisory Board (SAB) (2000*), Toward Integrated Environmental Decision-Making,* Washington.

Felt U., H. Nowotny and K. Taschwer (1995), *Wissenschaftsforschung. Eine Einführung,* Frankfurt a.M., New York: Campus.

Gellner, W. (1995), *Ideenagenturen für Politik und Öffentlichkeit: Think Tanks in den USA und in Deutschland,* Opladen: Westdeutscher Verlag.

Gibbons, M., C. Limoges, H. Nowotny, S. Schwartzman, P. Scott and M. Trow (1994), *The New Production of Knowledge. The Dynamics of Science and Research in Contemporary Societies,* London: Sage.

Gieryn, T.F. (1995), 'Boundaries of science', in S. Jasanoff, G. Markle, J. Petersen and T. Pinch (eds.), *Handbook of Science and Technology Studies,* London: Sage Publications, pp. 393–443.

Glynn, S., P. Cunningham and K. Flanagan (2001), *Science and Governance: Describing and Typifying the Scientific Advice Structure in the Policy Making Process – A Multi-National Study,* ESTO Poject Report, JRC Institute Prospective Technological Studies Sevilla.

Glynn, S., P. Cunningham and K. Flanagan (2003), *Typifying Scientific Advisory Structures and Scientific Advice Production Methodologies (TSAS),* Draft Final Report, University of Manchester.

Habermas, J. (1964), 'Verwissenschaftlichte Politik und öffentliche Meinung', in R. Reich and W. Bretscher (eds.), *Humanität und politische Verantwortung. Eine Beitragssammlung,* Stuttgart: Rentsch, pp. 104–20.

Halffman, W. (2003), *Science/Policy Boundaries: National Styles?',* discussion paper of the workshop *Auf dem Weg in die Wissensgesellschaft,* Institute of Science and Technology Studies, Bielefeld University, February 6–7, Bielefeld.

Halliwell, J.E., W. Smith and M. Walmsley (1999), *Scientific Advice in Government Decision-Making. The Canadian Experience. A Report in Support of the Work of the Council of Science and Technology Advisors,* Ontario, Canada: JEH Associates Inc.

Hammond, K.R., J.L. Mumpower, R. Dennis, S. Fitch and W. Crumpacker (1983), 'Fundamental obstacles to the use of scientific information in public policy making', *Technological Forecasting and Social Change* **24**: 287–97.

Heinrichs, H. (2002), *Politikberatung in der Wissensgesellschaft. Eine Analyse umweltpolitischer Beratungssysteme,* Wiesbaden: Deutscher Universitätsverlag.

Hilgartner, S. (2000), *Science on Stage: Expert Advice as Public Drama,* Stanford, CA: Stanford University Press.

Inglehardt, R. (1995), *Kultureller Umbruch,* Frankfurt a.M. and New York: Campus.

Jäger, W. and W. Welz (eds.), (1998), *Regierungssystem der USA,* München: R. Ouldenburg.

Jasanoff, S. (1990), *The Fifth Branch: Science Advisors as Policymakers,* Cambridge, MA: Harvard University Press.

Joss, S. and J. Durant (1995), *Public Participation in Science: The Role of Consensus Conferences in Europe,* London: The Science Museum/European Commission.

Kleimann, B. (1996), 'Das Dilemma mit den Experten – Ein Expertendilemma?', in H.-U. Nennen and D. Garbe (eds.), *Das Expertendilemma,* Berlin, Heidelberg, New York: Springer, pp. 183–215.

60 HARALD HEINRICHS

Knorr-Cetina, K. (1984), *Die Fabrikation von Erkenntnis. Zur Anthropologie der Wissenschaft*, Frankfurt a.M.: Suhrkamp.

Kreibich, R. (1986), *Die Wissenschaftsgesellschaft*, Frankfurt a.M.: Suhrkamp.

Krevert, P. (1993), *Funktionswandel der wissenschaftlichen Politikberatung in der BRD. Enwicklungslinien, Probleme und Perspektiven im Kooperationsfeld von Politik, Wissenschaft und Öffentlichkeit*, Münster: LIT Verlag.

Krimsky, S. (1984), 'Epistemic considerations on the value of folk-wisdom in science and technology', *Policy Studies Review* **3**, 2: 246–67.

Krimsky, S. (2000), *Hormonal Chaos. The Scientific and Social Origins of the Environmental Endocrine Hypothesis*, Baltimore, MD: Johns Hopkins University Press.

Lester, J.P. (ed.), (1995), *Environmental Politics and Policy: Theories and Evidence*, Durham, NC: Duke University Press.

Long, R.C. and T.C. Beierle (1999), *The Federal Advisory Committee Act and Public Participation in Environmental Policy*, Discussion Paper 99–17, Washington. Resources for the Future.

Luhmann, N. (1984), *Soziale Systeme. Grundriß einer allgemeinen Theorie*, Frankfurt a.M.: Suhrkamp.

Luhmann, N. (2000), *Die Politik der Gesellschaft*, Frankfurt a.M.: Suhrkamp.

Maasen, S. (1999), *Wissenssoziologie*, Bielefeld: Transcript Verlag.

Mannheim, K. (1995), *Ideologie und Utopie*, Frankfurt a.M.: Klostermann.

Martin, B. and E. Richards (1995), 'Scientific knowledge, controversy, and public decision making', in S. Jasanoff, G. Markle, J. Petersen and T. Pinch (eds.), *Handbook of Science and Technology Studies*, London: Sage Publications, pp. 506–27.

Mayntz, R. (1994), 'Politikberatung und politische Entscheidungsstrukturen: Zu den Voraussetzungen des Politikberatungsmodells', in A. Murswiek (ed.), *Regieren und Politikberatung*, Opladen: Westdeutscher Verlag, pp. 17–29.

Michaels, D., E. Bingham, L. Boden, R. Clapp, L.R. Goldman, P. Hoppin, S. Krimsky, C. Monforton, D. Ozonoff and A. Robbins (2002), 'Advice without dissent', *Science Magazine (editorial)*: 298.

Murswieck, A. (1994), *Regieren und Politikberatung*, Opladen: Westdeutscher Verlag.

Nelkin, D. (1979), 'Scientific knowledge, public policy, and democracy: A review essay', *Knowledge: Creation, Diffusion, Utilization* **1**: 106–22.

Nowotny, H. (1993), 'Experts and their expertise: On the changing relationship between experts and their public', *Bulletin of Science, Technology and Society* **1**: 235–41.

Nowotny, H. (1999), *Es ist so. Es könnte auch anders sein. Über das veränderte Verhältnis von Wissenschaft und Gesellschaft*, Frankfurt a.M.: Suhrkamp.

Nowotny, H., P. Scott and M. Gibbons (2001), *Re-Thinking Science. Knowledge an the Public in an Age of Uncertainty*, Cambridge, UK: Polity Press.

Oxford Economic Research Associates Ltd. (OXERA) (2000), *Policy, Risk and Science: Securing and Using Scientific Advice*, Oxford.

Parsons, T. (2000), *Das System moderner Gesellschaften*, Weinheim: Juventa.

Renn, O. (1995), 'Styles of using scientific expertise: A comparative framework', *Science and Public Policy* **22**, 3: 147–56.

Renn, O. (1999a), 'Sozialwissenschaftliche Politikberatung. Gesellschaftliche Anforderungen und gelebte Praxis', *Berliner Journal für Soziologie* **4**: 531–48.

Renn, O. (1999b), 'A model for an analytic-deliberative process in risk-management', *Environmental Science & Technology* **33**, 18: 3049–55.

Renn, O., T. Webler and P. Wiedemann (eds.), (1995), *Fairness and Competence in Citizen Participation*, Dordrecht: Kluwer Academic Publishers.

Rich, R.F. and C.H. Oh (2000), 'Rationality and use of information in policy decisions. A search for alternatives', *Science Communication* **22**, 2: 173–211.

Rip, A. (1985), 'Experts in public arenas', in H. Otway and M. Peltu (eds.), *Regulating Industrial Risks. Science, Hazards and Public Protection*, London: Butterworths, pp. 4–110.

Schimank, U. (1996), *Theorien gesellschaftlicher Differenzierung*, Opladen: Leske + Budrich.

Sebaldt, M. (1997), *Organisierter Pluralismus: Kräftefeld, Selbstverständnis und politische Arbeit deutscher Interessengruppen*, Opladen: Westdeutscher Verlag.

Smith, W. and J. Halliwell (1999), *Principles and Practices for Using Scientific Advice in Government Decision Making: International Best Practices*, Report to the S&T Strategy Directorate Industry, Canada.

Stehr, N. (1994), *Arbeit, Eigentum und Wissen: Zur Theorie von Wissensgesellschaften*, Frankfurt a.M.: Suhrkamp.

Stehr, N. (2003), *Wissenspolitik*, Frankfurt a.M.: Suhrkamp.

Weber, M. (1976), *Wirtschaft und Gesellschaft*, Tübingen: Mohr.

Weinberg, A. (1972), 'Science and trans-science', *Minerva* **10**, 2: 209–22.

Weingart, P. (1988), 'Verwissenschaftlichung der Gesellschaft – Politisierung der Wissenschaft', *Zeitschrift für Soziologie* **12**, 3: 225–41.

Weingart, P. (1999), 'Scientific expertise and political accountability: Paradoxes of science in politics', *Science and Public Policy* 26, 3: 151–61.

Weingart, P. (2001), *Die Stunde der Wahrheit. Zum Verhältnis der Wissenschaft zu Politik, Wirtschaft und Medien in der Wissensgesellschaft*. Weilerswist: Velbrück Wissenschaft.

Weiss, C.H. (1974), 'The circuitry of enlightenment. Diffusion of social science research to policymakers', *Knowledge: Creation, Diffusion, Utilization* **8**, 2: 274–81.

Wingens, M. (1989), *Soziologisches Wissen und politische Praxis. Neuere theoretische Entwicklungen der Verwendungsforschung*, Frankfurt a.M.: Suhrkamp.

Wynne, B. (1991), 'Sheep farming after Chernobyl: A case study in communicating scientific information', in B.V. Lewenstein (ed.), *When Science Meets the Public*, Washington, DC: American Association for the Advancement of Science.

Zilleßen H. (1993), 'Die Modernisierung der Demokratie im Zeichen der Umweltproblematik', in H. Zilleßen, P.C. Dienel and W. Strubelt (eds.), *Die Modernisierung der Demokratie*, Opladen: Westdeutscher Verlag, pp. 17–39.

CHAPTER 4

DAVID H. GUSTON

INSTITUTIONAL DESIGN FOR SOCIALLY ROBUST KNOWLEDGE: THE NATIONAL TOXICOLOGY PROGRAM'S REPORT ON CARCINOGENS

INTRODUCTION

The delegation of significant authority from political to scientific actors is arguably the central problem in science policy, both analytically and practically (Guston 1996). Varieties of the central problem of delegation play out through the logic of principal-agent theory, as described by an increasing amount of work that concentrates on questions of the sponsorship of research, the role of research councils, and other aspects of what Brooks (1968) famously called 'policy for science.'[1]

Yet, delegation is central not only to the patronage relationship but also to the advisory relationship. Principal-agent theory can therefore also help illuminate the structure of science policy with respect to questions of 'science in policy.' Such questions include issues of peer review and other aspects of the use of expert advice for making policy decisions. By framing the central problem of science policy as one of delegation, scholars gain perspective on the deceptively simple question that politicians or the public may ask, "How do we trust scientists when they say and do things we have little substantive knowledge about?"

This chapter draws on an in-depth case study (Guston 2003) of regulatory science in the United States – the creation and production of the biennial *Report on Carcinogens* by the National Toxicology Program (NTP) of the National Institute of Environmental Health Sciences. It uses principal-agent theory to make sense of the problems that the actors themselves faced in attempting to design a process to produce what scholars would call "socially robust knowledge" (Nowotny 2003). The first section of the chapter below briefly introduces relevant points of principal-agent theory to articulate a preliminary structure of 'science in policy.' The subsequent sections elaborate how these issues play out in the design of NTP's process for identifying carcinogens: the environment that precipitated Congress's need for a reliable agent; the creation of an intermediary to serve as that agent; the articulation of an explicit set of terms for the performance of that contract; and the avoidance of such rules that agents inevitably engage in. In the discussion and conclusion, I argue that this understanding of NTP's arrangements contributes to questions of institutional design by showing how NTP satisfies a variety of desiderata suggested in recent

63

Sabine Maasen and Peter Weingart (eds.), Democratization of Expertise? Exploring Novel Forms of Scientific Advice in Political Decision-Making – Sociology of the Sciences, vol. 24, 63–79.
© Springer Science+Business Media B.V. 2009

literature, particularly focusing on the benefits of voting over consensus as a method of expressing scientific judgment.

THE STRUCTURE OF 'SCIENCE IN POLICY'

There is a millennia-old perception that experts stand apart from, and superior to, ordinary people over whom they rightfully have authority. The philosophers of Plato's *Republic*, whose rule was underpinned by the Golden Lie, are prototypes of a variety of guardians of the commonweal that appear in the political theories of writers as diverse as Confucius, Lenin, and Skinner (Dahl 1989). Robert Dahl (1989) emphasizes in his critique of guardianship that guardians are a class of rulers to whom authority has been alienated, that is, yielded permanently and unaccountably, and he offers a variety of (surprisingly pragmatic) reasons why such guardianship should be rejected. First, there is no science of governing accessible only to a limited class of people and, even if there were, there would be no reasonable way to identify and train prospective guardians and secure their orderly transition. Additionally, no one guardian could possess the entirety of governing knowledge. Thus – and this point is under-appreciated – any committee of guardians would have to admit decision rules and other kinds of politics into their allegedly objective decision making.

Even if, however, we are freed from the specter of the alienated authority of guardianship, we still may be haunted by the troubles of delegated authority in which experts still rule with a practical if not actual lack of accountability. That there is no solution to the problem of accountability of experts in modern society is, for example, the worry of Stephen Turner (2003) in his recent *Liberal Democracy 3.0*. Asking a version of a question that has plagued pluralist thinkers, Turner asks of the role of experts, can liberal-democracies manage non-democratic sub-systems? I argue that we need not push the question as far as Turner has, and that we can still think of making expert sub-systems sufficiently democratic and accountable through appropriate institutional design.

Insights for this design come from principal-agent theory, used here as a heuristic device to speak somewhat more formally of a relatively ignorant principal who makes a delegation of authority to a relatively expert agent who receives that delegation. That the agent is more expert than the principal raises the prospect of two problems, known in the literature as adverse selection (or hidden information) and moral hazard (or hidden behavior). These problems are often understood by their temporal sequence. Adverse selection is the difficulty of choice the principal first faces in selecting the best agent to accomplish the chosen goals. The information that is hidden is precisely who is the best agent to delegate to or to fulfill the contract. Moral hazard is the difficulty the principal faces after the agent has been chosen and the contract let. The behavior that is hidden is how well the agent works to complete the delegation or to fulfill the contract.

Although the principal-agent literature is often about the control of the agent by the principal, both principals and agents have their own respective interests in the relationship, and these interests not only create the challenges of adverse selection and moral hazard, but they also contribute to sustaining a mutually beneficial relationship over time. Thus, as Sheila Jasanoff (2003a: 158) has asked, "since accountability is a two-way street, demanding not only a responsible agent but also a vigilant principal, how can decisionmaking procedures be designed to facilitate the public's supervisory role?"

'Science in policy' questions are primarily structured as problems of adverse selection. Decision makers have questions for which there may be technical answers, and they must choose which experts to believe among the many offering expertise. From a delegation or hypothetical contract between decision makers and experts, the former can receive benefits including: expert knowledge, insight, or early warning; the potential solution to particular problems or questions; and legitimation for decisions that require technical sophistication. The experts receive benefits including: direct payment as consultants or employees; indirect payment through appointments to positions that bestow authority, prestige, or access to specialized or privileged knowledge; and the psychic returns of seeing one's ideas implemented in a legitimated pursuit of the public good.

As initially conceived, this perspective appears to assume that decision makers are sincere in their desire to hear scientific perspectives and that experts are sincere in offering perspectives they believe are correct. Such an assumption, however, is not necessary because sincerity or the lack of it can be included in the framework of adverse selection. That is, some principals or some agents may decide to solve the problems of adverse selection by contracting only with others who are ideologically predisposed to agree. Indeed, this situation may be the prevailing norm of science advice.[2] One would then need to assume only that they want to transact with one another, and leave any speculation about the benefits from the transaction to observing the performance of the contract. Decision makers seeking only legitimation, for example, are likely to behave differently than those seeking early warning. Moreover, it is also plausible that many decision makers who appear insincere are merely overwhelmed by the problem of adverse selection. That is, they may behave as if they were not invested in sincere expert advice because the existing asymmetry of information has allowed insincere experts to convince them of their perspective. That disingenuous experts can deceive decision makers does not mean that decision makers do not desire sincere advice, although it may mean that decision makers can engage in facilitated self-delusion. But more generally it does mean that problems of agency are critical to public decision making.

Embedded in this discussion is a further assumption that the opinions of scientists can and actually do differ. If scientific consensus were truly monolithic, then although the asymmetry of information would still exist between politics and science broadly speaking, there would be no need to select among agents. Once a question was framed as a scientific one, one scientist's opinion would be just as good as the next. Although incorporating elements of adverse selection as well, the framing of 'scientific' or 'non-scientific' would displace the concerns attended to here. But because disagreement – even controversy – is a natural condition of the scientific community, at least three addition problems arise. First, even if closure might be anticipated, many political decisions cannot await eventualities, and decisions must be made in the absence of consensus or closure. Second, scientific consensus or closure is often a temporary phenomenon, ready to be overturned with the appearance of additional, compelling evidence. Third, scientific consensus or closure is not nor-

66 DAVID H. GUSTON

mally the product of entirely rational procedures, and neither is it the product of impersonal, market-like interactions. Such difficulties mean that political principals cannot rely on autonomously produced consensus or closure among scientific agents, but rather they must devise strategies of choice and delegate to chosen agents.[3]

As in the case of health insurance, those potential agents most actively seeking to join the contract may have the greatest propensity to incur costs for the principals, i.e., potential agents who will benefit most directly from the contract may provide self-serving information to the decision makers. One example is the problem of conflicts of interest among expert advisors. In most situations in the US, potential advisors must have a direct financial conflict, e.g., they must work for a company whose product will be regulated by the contemplated action, in order to be disqualified from participating in a regulatory science analysis. Another typical example of self-serving behavior is the recommendation for more research that experts often offer, even if more research does not reduce uncertainty, lead to greater consensus, or otherwise accord with the decision makers' aims by not actually being a necessary precondition for substantive political progress on the issue.[4]

One can derive a variety of strategies that a political principal would deploy in order to assure – that is, to attempt to overcome doubt about – the soundness of the delegation of authority implicit in the exercise of scientific judgment for policy making. In the case of providing health insurance, the typical strategies to resolve problems of adverse selection involve excluding from the contract any potential agents who are or have a propensity to be ill, and providing incentives for those agents who do become party to the contract to remain healthy. The former solution typically requires the use of a monitor or intermediary, e.g., a physician who will examine potential agents for pre-existing or excluded conditions. This solution, however, raises that timeless, reiterative problem: Who will watch the watcher? The latter solution requires writing a contract with appropriately detailed terms, e.g., discounts for completing fitness courses, although such solutions impose analytical costs in calculating the incentives and adjusting the terms of the contract properly. Nevertheless, principals engage in such strategies of mediation and detailed procedures in their attempts to assure the integrity of the delegation of authority.

The remainder of this chapter applies this nascent framework to help order a case of 'science in policy' in the United States. It frames the discussion around mediation and explicit procedures as implemented by the National Toxicology Program (NTP), a small agency in the US National Institute of Environmental Health Sciences (NIEHS) – which is itself one of the more than two dozen National Institutes of Health. These strategies include NTP's intermediation between politicians and scientific agents, the writing of explicit contracts governing the behavior of those agents, and the promulgation of various rules that make the behavior of the scientific agents more observable. These strategies help to produce socially robust knowledge but they are not, however, perfect, and the scientific agents do in fact find ways to 'shirk.'

THE CASE OF SACCHARIN, PART 1: NEED FOR A RELIABLE AGENT

Saccharin, a derivative of coal tar, has a long and controversial history as a nonnutritive sweetener and food additive (Priebe and Kauffman 1980; Cummings 1986;

Marcus 1997). After Congress passed the Food Additive Amendments of 1958 to the Food, Drug and Cosmetic Act, the scientific and regulatory communities considered saccharin "generally recognized as safe" (GRAS). Subsequent experimental evidence gathered in the 1960s, however, led the Food and Drug Administration (FDA) to revoke saccharin's recognition as safe in February 1972.

FDA also issued an interim guideline forbidding any new uses for saccharin while it awaited a report from the National Academy of Sciences (NAS). The interim guidelines were set to expire at the end of June 1973, but FDA extended them indefinitely, citing studies that found significant increases in the incidence of bladder cancer in the male offspring of test animals fed saccharin (U.S. Senate 1977: 23). In December 1974, NAS submitted its review of the various studies, suggesting that saccharin was a carcinogen, but pointing to serious problems in the studies because the effective agent could have been impurities rather than the saccharin itself. In Canada, a study was designed to resolve this ambiguity, but Senator Gaylord Nelson (Democrat-Wisconsin), chairman of the Select Committee on Small Business, thought FDA was dawdling, and he asked the General Accounting Office (GAO) to investigate FDA's handling of the regulation of food additives (Marcus 1997). In testimony before Nelson's committee in January 1977, GAO critiqued FDA's regulation of saccharin and "recommended that [FDA] promptly reassess ... the need for ... possibly discontinuing [saccharin's] use in food" (U.S. Senate 1977: 27). Shortly thereafter, the Canadian study found that saccharin, rather than the impurities, caused bladder cancer in rats. Invoking the Delaney clause – a provision in the 1958 Amendments that prohibited any carcinogens from being added to foods – FDA proposed in the *Federal Register* on 15 April 1977 to ban saccharin.

The public reacted to the proposed ban with an outcry over losing the last substitute for sugar, as cyclamate had been banned in the 1960s. Congress responded, in part, by requesting a report from the Office of Technology Assessment. OTA surveyed the available scientific evidence on the carcinogenicity of saccharin, explored its potential health benefits for some consumers, and – in an unusual move for the policy analytic organization – commissioned Ames tests of saccharin's potential mutagenicity. OTA (1977: 5f.) concluded that "[l]aboratory evidence demonstrates that saccharin is a carcinogen," albeit a weak one, and one for which epidemiological studies had not shown a carcinogenic effect in humans. Nevertheless, saccharin seemed to meet the criteria proposed by the Occupational Safety and Health Administration to identify a 'confirmed' carcinogen. Not wanting to completely disregard FDA and the Delaney clause, Congress passed the Saccharin Study and Labeling Act (P.L. 95–203), which placed a moratorium on the saccharin ban, required labeling of all food products containing saccharin, and directed NAS to study the issue further.

Congress could not abide, however, such a sloppy, dilatory process whenever some scientists suspected a potential carcinogen in the food supply. OTA, NAS, and FDA, as well as private sector interests both for and against the continued use of saccharin, had a stake in assessing its carcinogenicity. Which agent should Congress choose: FDA, which applied a troublesome legal standard literally? NAS, which hemmed and hawed and asked for more research? OTA, which confirmed saccharin's mutagenicity but balked on the epidemiology? Industry or patient groups with significant commercial and other interests to protect? Without consensus in the scien-

68   DAVID H. GUSTON

tific community, and with the presence of patently self-interested advocates, Congress needed a reliable agent to identify carcinogens in future conflicts.

CREATING NTP AND THE *REPORT ON CARCINOGENS*

Not quite one year after it instructed FDA to defer regulatory action on saccharin, Congress passed the Biomedical Research Extension Act (P.L. 95–622) which, among other provisions, required the Secretary of the Department of Health, Education and Welfare (DHEW; now the Department of Health and Human Services, DHHS) to publish an annual report listing substances known or anticipated to be human carcinogens. Congress mandated that DHEW perform the task but delegated the design of a process that would fulfill the mandate. In 1979, DHEW created the National Toxicology Program (NTP) to implement the mandate by publishing a *Report on Carcinogens*.

NTP established an elaborate advisory system to identify human carcinogens. Initially, two review groups, the NIEHS/NTP Review Committee (RG1) and the NTP Executive Committee's Interagency Working Group (RG2), contributed to the *Report*'s decision making. In the first step of a detailed and iterative process, NTP receives a petition from any individual or group nominating a substance for consideration.[5] NTP then solicits public comment through notification in the *Federal Register*, trade journals, and its own publications.[6] RG1 receives the original petition and all public comments and decides if the substance warrants further consideration. If not, the petition is returned to the petitioner, who can resubmit it with further justification. Otherwise, RG1 appoints a primary and secondary reviewer from within its ranks to shepherd the petition through the committee. The primary reviewer identifies relevant articles from only the peer-reviewed literature and, with the assistance of the secondary reviewer, selects those articles to be included in a draft report, which is prepared by staff with the assistance of a contractor.[7] The reviewers then examine the petition, the citations, and the draft report for completeness and accuracy and, after making any necessary revisions, the primary reviewer presents it to RG1. RG1 considers this material, as well as the public comments in response to the petition, and makes a recommendation for listing or delisting. RG1 can also conclude that, after the review, there is still insufficient information and return the petition to the petitioner. The members of RG1 vote on the recommendation, and RG1 forwards the petition to RG2.

RG2 receives the petition, the public comments, and the draft report. It assigns another reviewer, who leads RG2's iteration of roughly the same procedure. RG2 provides comments and recommendations for any changes to the draft report and votes its recommendation for listing or delisting the substance. In the initial design of the process, NTP's Executive Committee would then review the entire record for each substance, vote on each substance, and forward the record with its own comments and voting results to the director of NTP for decision. NTP would then submit the *Report* to the Secretary for review and approval and, finally, to Congress for publication.

Through this process, NTP institutionalized the determination of what substances are or might reasonably be anticipated to be human carcinogens. Through the review

INSTITUTIONAL DESIGN FOR SOCIALLY ROBUST KNOWLEDGE 69

groups, NTP gathered many of the various experts from agencies that might otherwise have offered opinions directly themselves, and it solicited public input in a coherent and informed way. NTP became the analogue of the physician, the agent of Congress who is intermediary to other potential agents who were themselves attempting to assess the carcinogenicity of substances more directly.

This process embodied several strategies to combat the problems of agency. Limiting the agents to government employees minimized the threat of conflicts of interest, as did limiting the information used to the peer reviewed literature. Although NTP sought public comment, no advocates and no information produced purely for advocacy could be dispositive in its decisions. The creation of two advisory committees, with two different constituencies, increased the amount of information produced for the principal.[8] Furthermore, the recommendations of the advisory committees are exactly that – recommendations. Political principals, including the NTP director, the department Secretary, and Congress itself are responsible for the listing or delisting of a substance. This authority is more substantive than simply formal because of the relatively obvious fact that RG1 and RG2 may sometimes disagree. The NTP director must then decide how to cope with a substance despite divergent expert assessments. Such was the case with the decision about saccharin, described further below.

TOWARD A MORE EXPLICIT CONTRACT

NTP followed this procedure until the early 1990s. In the 1993 NIH Revitalization Act (P.L. 103–46), Congress mandated biennial rather than annual reports to provide "timely and useful scientific information to the regulatory agencies and the public while providing savings that would be better spent on testing additional agents" (US Senate 1992: 41). NTP published the first biennial report, and the eighth overall, in 1998 (NTP 1998). The *Eighth Report* also implemented two crucial changes NTP made in 1995: the creation of a new, more public advisory committee, and the revision of the criteria used for listing carcinogens.

Through the first change, NTP expanded its review procedures by adding a third committee – a standing subcommittee of the Board of Scientific Counselors (DHHS 1996). Like RG1 and RG2, this *Report on Carcinogens* Subcommittee appoints a reviewer from within its ranks to guide its discussion about a nominated substance. Unlike RG1 and RG2, however, the *Report on Carcinogens* Subcommittee deliberates in public. Because it comprises members who are not all employees of the federal government, the Subcommittee falls under the jurisdiction of the Federal Advisory Committee Act (FACA, P.L. 92–463). In addition to mandating public meetings, FACA requires that such committees be 'fairly balanced' in their composition. NTP announces meetings of the Subcommittee in the *Federal Register* and other publications, soliciting groups or individuals to submit written comments or to address the Subcommittee during its public meeting. Based on the prior record and any relevant public comment, the Subcommittee makes further recommendations for changes to the draft document and votes on a recommendation for listing or delisting.

By creating this committee of external advisors, NTP invited a greater risk of conflicts of interest, although FACA ameliorates the worst kinds. NTP also extended the logic of multiple advisory panels to release more information for the principal by including non-governmental experts from industry, academe, and labor.[9]

In 1995, NTP also changed the criteria through which the various advisory committees arrive at their conclusions. As mentioned above, the committees deliberate on four possible outcomes for any nominated substance: the information is insufficient for deliberation; it should be listed as a known human carcinogen; it should be listed as reasonably anticipated to be a human carcinogen; or it should not be listed or be delisted. The original criteria maintained that a substance should be listed as a known human carcinogen if and only if "[t]here is sufficient evidence of carcinogenicity from studies in humans that indicates a causal relationship between the agent and human cancer" (DHHS 1995: 30435). The original criteria maintained that a substance should be listed as reasonably anticipated to be a human carcinogen if and only if:

a. There is limited evidence of carcinogenicity from studies in humans, which indicates that causal interpretation is credible, but that alternative explanations, such as chance, bias or confounding, could not adequately be excluded, or

b. There is sufficient evidence of carcinogenicity from studies in experimental animals that indicates that there is an increased incidence of malignant tumors: (a) in multiple species or strains, or (b) in multiple experiments (preferably with different routes of administration or using different dose levels), or (c) to an unusual degree with regard to incidence, site or type of tumor or age at onset. Additional evidence may be provided by data concerning dose-response effects, as well as information on mutagenicity or chemical structure (DHHS 1995: 30435).

The criteria make precise science policy statements about how the agents are supposed to handle evidence, e.g., they must rely on 'increased incidence of malignant tumors' and not, for example, consider benign tumors or lesions.[10] They also specify the inadequacy of a single animal-system model of carcinogenicity under normal circumstances.

In April 1995, the Board of Scientific Counselors created an ad hoc working group, which held a public meeting to consider revising the listing criteria and procedures (DHHS 1995).[11] The working group did not recommend any changes to the criteria for determining a known human carcinogen. The stated criterion was modified in a modest way to instruct for a finding of known human carcinogenicity when "[t]here is sufficient evidence of carcinogenicity from studies in humans that indicates a causal relationship between *exposure to* the agent, *substance or mixture* and human cancer" (DHHS 1996; changes noted in italics).

The working group did, however, recommend substantive changes to the criteria governing the finding that a substance is reasonably anticipated to be a human carcinogen. The proposed criteria included consideration of route of exposure, mechanisms of action, and sensitive subpopulations. NTP adopted these suggestions, expanding them to include membership in a "well defined, structurally-related class of substances whose members are listed in a previous Annual or Biennial Report on Carcinogens ... or there is convincing relevant information that the agent acts through mechanisms indicating it would likely cause cancer in humans." NTP also added a descriptive paragraph:

> Conclusions regarding carcinogenicity in humans or experimental animals are based on scientific judgment, with consideration given to all relevant information. Relevant information includes, but is not limited to, dose response, route of exposure, chemical structure, metabolism, pharmacokinetics, sensitive sub populations, genetic effects, or other data relating to the mechanism of action or factors that may be unique to a given substance. For example, there may be substances for which there is evidence of carcinogenicity in laboratory animals but there are compelling data indicating that the agent acts through mechanisms which do not operate in humans and would therefore reasonably be anticipated not to cause cancer in humans (DHHS 1996: 50499).

Congress did not impose these specific controls. Rather, the intermediary developed them under the principal's watchful eye in order for the agents to demonstrate their successful performance of the delegation. No one would complain to Congress about NTP if its procedures were more open to FACA, whose requirements for openness as well as balance combat adverse selection.[12] Scientists would not feel abused if NTP's criteria were made more explicit and attuned to 'scientific judgment.' Indeed, NTP director Kenneth Olden held that the new criteria and processes provided for "better science and better responsiveness" (DHHS 1998).

THE CASE OF SACCHARIN, PART II: SHIRKING BEHAVIOR

NTP first listed saccharin as reasonably anticipated to be a human carcinogen in its *Second Report*, published in 1981, and saccharin appeared in all subsequent reports up to and including the *Eighth Report*. Responding to the call for nominations for the *Ninth Report*, the Calorie Control Council (1997) nominated saccharin for delisting "on the basis of NTP's new criteria incorporating the use of mechanistic data."[13] By the end of September 1997, NTP had completed the draft background document on saccharin. In addition to reviewing toxicological and epidemiological studies of saccharin, the draft argued that:

> [t]here is evidence of the carcinogenicity of saccharin in rats but less convincing evidence in mice. Mechanistic studies indicate that ... [t]he factors thought to contribute to tumor induction by sodium saccharin in rats would not be expected to occur in humans. The mouse data are inconsistent and require verification by additional studies. Results of several epidemiological studies indicate no clear association between saccharin consumption and urinary bladder cancer. Although it is impossible to absolutely conclude that it poses no threat to human health, sodium saccharin is not reasonably anticipated to be human carcinogen under conditions of general usage as an artificial sweetener (NTP 1997: RC3).

The draft report thus argued that the criterion of multiple sites or species was not fulfilled. RG1 and RG2 voted 7-3 and 6-2, respectively, to delist saccharin (DHHS 1998y), setting off speculation in the press about saccharin's ultimate absolution (e.g., Huber 1997; Kaiser 1997; McGinley 1997).

72 DAVID H. GUSTON

Table 1: Votes of the members of the Report on Carcinogens *subcommittee on saccharin for the* Ninth Report

Voting to Delist	Voting to Retain Listing
A. John **Bailer**, Ph.D. Department of Mathematics & Statistics Miami University	Eula **Bingham**, Ph.D. Departmentt of Environmental Health University of Cincinnati College of Medicine
Steven A. **Belinsky**, Ph.D. Inhalation Toxicology Research Institute Kirland Air Force Base	George **Friedman-Jimenez**, M.D. School of Public Health Bellevue Hospital
Clay **Frederick**, Ph.D. Mechanistic Toxicology Group Rohm and Haas Company	Nicholas K. **Hooper**, Ph.D. Department of Toxic Substances Control California Department of Health Services
	Franklin E. **Mirer**, Ph.D. Health and Safety Department UAW International

Not Voting:

Arnold L. **Brown**, M.D. University of Wisconsin Medical School (Chair; only votes in case of tie)	Carol J. **Henry**, Ph.D. Health Environmental Science Department American Petroleum Institute (absent)

The *Report on Carcinogens* Subcommittee held its public meeting on 30–31 October 1997 to review the recommendations of RG1 and RG2 and hear additional public comment – offered by the Calorie Control Council in favor of delisting and by the Center for Science in the Public Interest opposed to delisting. The Subcommittee then voted 4-3 to reject the draft report and continue listing saccharin. Table 1 shows the members of the Subcommittee and how they voted.

Press reports suggest that the members of the Subcommittee who voted to retain saccharin on the list displayed a certain precautionary outlook that was outside the scope of the criteria. Nicholas K. ('Kim') Hooper from the California Department of Health Services said, "Delisting is going to weigh on my conscience if I'm wrong" (quoted in Stolberg 1997: A13). Franklin Mirer, the director of health and safety from the United Auto Workers, International (UAW), found the 'equivocal' data from human epidemiological studies – which showed an increased cancer risk among some subpopulations – reason enough not to delist: "What I'm saying is the epidemiology is perhaps not strong enough to identify saccharin as a carcinogen, but it doesn't rule out that it's a risk" (quoted in McGinly 1997). *The Wall Street Journal* reported that "[a]t least one member of the panel who voted to keep saccharin listed said he probably wouldn't have voted to add saccharin, if that had been the issue, but wasn't comfortable about delisting it" (McGinly 1997).

In confidential interviews with the author, members of the *Report on Carcinogens* Subcommittee diverged in their explanations of the outcome as much as they did in the voting itself. Some attributed the lack of consensus to individually different perspectives on risk-taking. Others attributed it to disciplinary differences. "I suspect that my particular bias," said a Subcommittee member, is "when in doubt, regulate." Still others attributed differences to political agendas, as some "people were determined to delist for political or science policy reason" and some "are very industry oriented and are hesitant to call something a carcinogen, especially when it is on the cusp." Indeed, the Subcommittee members who voted to retain saccharin's status as reasonably anticipated to be a human carcinogen developed a perspective about shirking that contradicts that portrayed in the media coverage, arguing that those seeking to delist saccharin in essence 'nullified the criteria.' Those favoring delisting overemphasized the mechanistic data, which did not logically eliminate mechanisms that could cause cancer in humans. They neglected evidence in female rats that may have contradicted the mechanistic data. They conflated the hazard identification task of NTP – which is simply to determine carcinogenic potential – with a risk assessment for human consumption, which no one believes is high for saccharin. They over-emphasized the worth of human epidemiological data because many cancers that saccharin might cause would not yet have shown up in the study populations.

After the vote of the *Report on Carcinogens* Subcommittee, both the UAW's Mirer and the chairman of the Subcommittee, Arnold Brown of the University of Wisconsin Medical School (who, as chairman, did not vote) "thought that it might be difficult" for NTP director Olden to contradict the panel and delist saccharin. But in December 1998, the full Executive Committee voted 6-3 to delist saccharin.

According to members of the Subcommittee, the mixed vote 'sends a message' about the underlying uncertainty in the data and the conflict of scientific judgment that the advisory committees could not have sent had they operated by consensus rather than reporting votes. The individual votes, and the record among the advisory committees, "should show the level of agreement that the data show," and the record of disagreement "alerts people to the fact that [different opinions were] considered." The full committee's vote meant that saccharin had, like a tennis player, won its match for delisting 7-3, 6-2, 3-4, 6-3. The *Ninth Report on Carcinogens*, finally issued in May 2000, contained 218 entries for substances known and reasonably anticipated to be human carcinogens (NTP 2000). Saccharin was not among them.

DISCUSSION

NTP was established under a delegation from Congress to assure the sound production of information about known and reasonably anticipated human carcinogens. As such, it is an expert agent for Congress, like the physician-intermediary between the health insurer and insurance seekers. NTP is a watcher of other agents – scientists themselves – who deliberate about what substances are or are not carcinogenic.

Recognizing, as Jasanoff (2003a: 159; emphasis in the original) writes, that "[e]xpertise is not so much *found* as *made* in the process of litigation or other forms of technical decisionmaking," NTP made a well-regulated scientific marketplace – what Nowotny (2003) might call an *agora* – in which some degree of consensus and closure, as well as the liberation of a good deal of information, could be expected. Congress did not mandate the architecture of this *agora* but NTP designed its procedures to demonstrate its faithful performance of the delegation. NTP established multiple advisory committees to represent interests both internal and external to the government. FACA protected the integrity of the input from external advisors against such threats of adverse selection as conflicts of interest. NTP promulgated specific science policy criteria, upon which it instructed the members of these advisory committees to formulate their judgments. NTP relied on voting, rather than consensus, to embody the uncertainty in the underlying data and communicate this additional information to political principals.

NTP created a process that also embodies Nowotny's (2003: 155) three characteristics of socially robust knowledge. First, NTP tests knowledge about carcinogenic potential "in a world in which social, economic, cultural and political factors shape the products and processes resulting from scientific and technological innovation." FACA and the open hearings of the *Report on Carcinogens* Subcommittee assured this after the 1995 procedural changes. Second, NTP 'extends' expertise throughout society in a similar way – not only by validating the participation of diverse interests through diverse committees and FACA mandates, but by allowing public comment to initiate scrutiny of substances, by soliciting public comment at all stages of its deliberations, and by preserving the discretion of a political appointee to make the final determination. NTP also crucially relies on science policy decisions – the rules under which individual experts operate – which are open to greater public scrutiny and influence than are the decisions about carcinogenic potential themselves. Third, NTP provides a forum for claims about carcinogenic potential to be 'repeatedly tested, expanded and modified.' New research and new rules created a situation in which saccharin was delisted. New research could create a situation in which saccharin could be relisted. NTP's *Report on Carcinogens* process seems to be an example, again quoting Jasanoff (2003a: 161), in which "a bounded but candid deliberation among the holders of divergent viewpoints could lead to ... a more accountable exercise of judgment, and eventually a better analysis."

Special attention should also be paid to the nature of voting in NTP's *agora*. Although, as realists often assert, one cannot repeal the law of gravity by voting, voting occurs more than is generally recognized in a variety of traditionally technical venues (Guston, under review). Balloting in this *agora* does not absolutely determine what substances are or are not human carcinogens, for a political actor still makes that specific determination, but the voting here certainly more than hints at the outcome.

I want to suggest that voting serves a number of specific functions, beyond this hinting. First, in the context of principal-agent theory, voting is a preferable method of aggregating the preferences of the participants because it liberates more information than does consensus, through which the agent speaks with only one voice.

Table 2: Votes of the members of the Report on Carcinogens *subcommittee on all substances, in comparison to the majority*

Name	More	As	Less
Bailer	1	21	1
Belinsky	0	18	5
Bingham	2	9	1
Frederick	1	18	2
Friedman-Jimenez	0	17	0
Henry	0	7	3
Hooper	3	19	1
Mirer	3	18	0
Hecht	0	12	0
Kelsey	1	14	0
Medinsky	0	8	2
Russo	1	7	2
Zahm	0	13	0

Second, voting assists in accountability because, in conjunction with rules on openness, voting connects individuals to their stances. Thus, the previous Table 1 can be replicated for every substance on which the *Report on Carcinogens* Subcommittee votes, and both analysts and the public can see how individual members of the Subcommittee vote. Aggregating the votes in particular ways, for example, by the sectoral or disciplinary affiliation of the Subcommittee member, can provide additional information about the balance of the Subcommittee under FACA. In Table 2, 'more' represents how many votes the individual cast that were 'more protective' than the majority of Subcommittee members cast for any substance, 'as' means how many votes were 'as protective' as the majority, and 'less' means how many votes were 'less protective' than the majority. A more protective vote would be voting to list as substance as 'reasonably anticipated to be a human carcinogen' when the majority voted not to list the substance, or voting to list it as a 'known human carcinogen'

DAVID H. GUSTON

when the majority voted to list it as 'reasonably anticipated.' A 'less protective' vote would be the other way around.[14]

Table 3 sums the votes by sectoral and disciplinary affiliation. One can readily note that the industry members are less protective overall than other members, and that the single labor member does not 'balance' out the other industry representatives, thus providing some empirical evidence about the satisfaction of FACA. In the disciplinary analysis, Subcommittee members affiliated with laboratory disciplines (e.g., toxicology) were less protective, those affiliated with populations and statistics (e.g., bio-statistics, epidemiology) were right in the middle, and those affiliated with organismal studies (zoology, medical doctor) were more protective. Laboratory disciplines were also more frequently represented than either of the other two.

Table 3: Votes of the members of the Report on Carcinogens *subcommittee on all substances, aggregated by sector and disciplinary group*

Sector/ Disciplinary Group	More	As	Less
Academic	5	80	4
Government	3	50	6
Industry	1	33	7
Labor	3	18	0
Stats/Pop	1	34	1
Organismal	7	65	3
Laboratory	4	82	13

That one can perform this admittedly crude but still potentially revealing analysis suggests a third reason to commend voting, as its analysis may open the door to a different kind of politics around such committees and around FACA – one that encourages empirical inquiry relevant to the selection of such committees.

CONCLUSION

After examining NTP's *Report on Carcinogens*, several levels of conclusions can be offered. The first concerns the framing of the case by principal-agent theory, which proves a handy map for issues of 'science in policy,' helping demonstrate how a political principal delegates authority to a scientific agent and how that agent adopts strategies to demonstrate its fulfilling the delegation in a competent way. Second, the chapter suggests that by focusing on the structure of delegation, relationships that meet reasonable normative criteria about expertise can be met. That is, through the appropriate design of institutions the production of socially robust knowledge can be successfully delegated rather than alienated. Such design elements include balanced participation from interested experts (as provided by FACA), clear science policy rules about how to come to scientific judgments, multiple sources of advice operating under similar rules and information, and open and transparent voting rules for expressing the scope of agreement of scientific judgment. Together, these elements provide both the democratization of expertise and the expertizing of democracy that Libertore and Funtowicz (2003) have called for.

Third, these design elements improve the conditions for accountability by teasing apart what Jasanoff (2003b) has identified as the 'three bodies of expertise': the individual experts themselves, the bodies of knowledge on which they draw, and the advisory bodies they constitute. The ability of experts to cloak their authority by speaking from a position of consensus, determined by unspecified procedures, prevents the differentiation, specification, or identification of responsibility that is needed for accountability. By designing institutions to provide expert advice according to these elements, we may be able to stave off asking Turner's unanswerable question about the compatibility of liberal-democratic governance with authority alienated to experts.

Bloustein School of Planning and Public Policy, Rutgers State University of New Jersey, New Brunswick, USA

NOTES

[1] This work includes Braun (1993, 1998), Braun and Guston (2003), Caswill (1998), Guston (1996, 1999, 2000), Morris (2000) and van der Meulen (1998).

[2] It is certainly what US Representative Henry Waxman (Democrat, California) believes is the norm of the Bush Administration, as Waxman released a report purporting to document dozens of episodes of the inappropriate politicization of science (US House of Representatives 2003).

[3] This argument is similar to that in Guston (2000) in which the political principal cannot rely on the autonomously produced integrity or productivity of the scientific agent and must therefore create new institutions to assure these requisites.

[4] Some believe this to be the case, for example, in the research agenda for climate change (e.g., Pielke and Sarewitz 2002).

[5] NTP may consider an "agent, substance, mixture, or exposure circumstance," but I will simply refer to "substance."

[6] This account is derived from NTP (1998), appendix C.

[7] This process of writing and reviewing the report is similar to the preparation of the criteria document for the National Ambient Air Quality Standards under the Clean Air Act Amendments of 1977 (see Jasanoff 1990: 102)

[8] Information from the confidential interviews supports this perspective, as informants distinguished between RG1 as an internal organ of NTP more concerned with toxicological evidence and RG2 as a broader, higher level committee more concerned with the political and regulatory consequences of decisions.

[9] In interviews, members of the Subcommittee supported this interpretation, distinguishing the Subcommittee from RG1 and RG2 by its public (as opposed to bureaucratic) constituency and its greater expertise in epidemiology, public health, and human exposure.

[10] See Jasanoff (1990) for documentation of conflicts in regulatory science committees over exactly such science policy issues.

78

DAVID H. GUSTON

[11] Another such meeting occurred 27–28 January 2004.

[12] This is one of the lessons from the literature on "fire-alarm oversight" by Congress over executive agencies (McCubbins and Schwartz 1987 [1984]).

[13] The Calorie Control Council represents the low-calorie and reduced-fat food and beverage industry. See http://www.caloriecontrol.org. In January 1997, FDA revoked a rule prescribing the display of warning signs at retail establishments about the sale of saccharin (DHHS, 1997z); FDA initiated the action following a petition from the Calorie Control Council and under authority of a bill to amend the Federal Food, Drug, and Cosmetic Act to repeal the saccharin notice requirement (P.L. 104-124) (DHHS 1996y).

[14] Each individual does not have the same number of votes because some may have joined the committee at different times in its deliberations, some may have missed meetings, and some may have abstained or declared conflicts of interest. All votes, however, are on substances considered for the ninth report.

REFERENCES

Braun, D. (1993), 'Who governs intermediary organizations? Principal-agent relations in research policy-making', *Journal of Public Policy* **13**, 2: 135–62.

Braun, D. (1998), 'The role of funding agencies in the cognitive development of science', *Research Policy* **27**: 807–21.

Braun, D. and D.H. Guston (eds.), (2003), 'Principal-agent theory and science policy – Special Issue', *Science and Public Policy*, **30**, 4.

Brooks, H. (1968), *The Government of Science*, Cambridge, MA: MIT Press.

Calorie Control Council (1997), 'Saccharin: Comments of the Calorie Control Council for the National Toxicology Program Board of Scientific Counselors' Meeting', Mimeo and enclosure to letter from Lyn O'Brien Nabors, Executive Vice President, Calorie Control Council, to Dr. Larry Hart, Executive Secretary, National Toxicology Program, October 24.

Caswill, C. (1998), 'Social science policy: Challenges, interactions, principals and agents', *Science and Public Policy* **25**, 5: 286–96.

Cummings, L.C. (1986), 'The political realities of artificial sweeteners', in H.M. Sapolsky (ed.), *Consuming Fears: The Politics of Product Risks*, New York: Basic Books, pp. 116–40.

Dahl, R. (1998), *Democracy and Its Critics*, New Haven, CT: Yale University Press.

Department of Health and Human Services (DHHS) (1995), 'Proposed revised criteria for listing substances in the Biennial Report on Carcinogens (BRC) and notice of meeting of the NTP Board of Scientific Counselors', *Federal Register* **60**, 110: 30433–6.

Department of Health and Human Services (DHHS) (1996y), 'Food labeling: Saccharin and its salts; retail establishment notice; revocation', *Federal Register* **61**, 189: 50770–71.

Department of Health and Human Services (DHHS) (1996), 'Notice: Revised criteria and process for listing substances in the biennial Report on Carcinogens', *Federal Register* **61**, 188: 50499-500.

Department of Health and Human Services (DHHS) (1997z), 'Food labeling: Saccharin and its salts; retail establishment notice', *Federal Register* **61**, 17: 3791-92.

Department of Health and Human Services (DHHS) (1998y), 'Call for public comments: Agents, substances, mixtures and exposure circumstances proposed for listing in or removing from the Report on Carcinogens, ninth edition', *Federal Register* **63**, 53: 13418–21.

Department of Health and Human Services (DHHS) (1998), 'Updated criteria approved for "anticipated" human carcinogens', *HHS News* (Press Release).

Guston, D.H. (1996), 'Principal-agent theory and the structure of science policy', *Science and Public Policy* **23**, 4: 229–40.

Guston, D.H. (1999), 'Stabilizing the boundary between US politics and science: The role of the Office of Technology Transfer as a boundary organization', *Social Studies of Science* **29**, 1: 87–112.

Guston, D.H. (2000), *Between Politics and Science: Assuring the Integrity and Productivity of Research*, New York: Cambridge University Press.

Guston, D.H. (2003), 'Principal-agent theory and the structure of science policy, revisited: "Science in policy" and the US Report on Carcinogens', *Science and Public Policy* **30**, 4: 347–57.

Guston, D.H. (under review), 'Parliamentary procedures for technology as legislation', in P. Hamlett and E.J. Woodhouse (eds.), *Since Autonomous Technology: Understanding Technics and Politics*, Cambridge, MA: MIT Press.

Huber, P. (1997), 'The health scare industry', *Forbes Magazine* (6 October): 114.

Jasanoff, S. (1990), *The Fifth Branch: Science Advisors as Policy Makers*, Cambridge, MA: Harvard University Press.

Jasanoff, S. (2003a). '(No?) Accounting for expertise', *Science and Public Policy* **30**, 3: 157–62.

Jasanoff, S. (2003b). 'Bodies of expertise', presented at the annual meeting of the Society for the Social Studies of Science, Atlanta, GA, 18 October.

Kaiser, J. (1997), 'Smoldering battle over Saccharin heats up', *Science* **278**: 791.

Libertore, A. and S. Funtowicz (2003), 'Introduction: "Democratising" expertise, "expertising" democracy: What does this mean, and why bother?' *Science and Public Policy* **30**, 3: 146–50.

Marcus, A.I. (1997), 'Sweets for the sweet: Saccharin, knowledge, and the contemporary regulatory nexus', *Journal of Policy History* **9**, 1 :33–47.

McCubbins, M.D. and T. Schwartz (1987 [1984]), 'Congressional oversight overlooked: Police patrols versus fire alarms', in M.D. McCubbins and T. Sullivan (eds.), *Congress: Structure and Policy*, New York: Cambridge University Press, pp. 426–40.

McGinley, L. (1997), 'How sweet it is: Saccharin's presence on list of carcinogens may be near an end', *The Wall Street Journal* (29 October): 1.

Meulen, B. van der (1998), 'Science policies as principal-agent games: Institutionalization and path-dependency in the relation between government and science', *Research Policy* **27**, 4: 397–414.

Morris, N. (2000), 'Science policy in action: Policy and the researcher', *Minerva* **38**: 425–51.

National Toxicology Program (NTP) (1997), *Draft RC Background Document for Saccharin* (September 25), Research Triangle, NC: NTP.

National Toxicology Program (NTP) (1998), *Eighth Report on Carcinogens*, [Research Triangle, NC: National Institute of Environmental Health Sciences].

National Toxicology Program (NTP) (2000), *Ninth Report on Carcinogens*, Research Triangle, NC: National Institute of Environmental Health Sciences.

Nowotny, H. (2003), 'Democratising expertise and socially robust knowledge', *Science and Public Policy* **30**, 3: 151-6.

Office of Technology Assessment (OTA) (1977, October), *Cancer Testing Technology and Saccharin*, NTIS Order No. PB-273499, Washington, DC: U.S. Government Printing Office, accessed at <http://www.wws.princeton.edu/cgi-bin/byteserv.prl/~ota/disk3/1977/7702/770201.PDF>.

Pielke, R., Jr. and D. Sarewitz (2002), 'Wanted: Scientific leadership on climate', *Issues in Science and Technology* **19**, 2: 27–30.

Priebe, P.M. and G.B. Kauffman (1980), 'Making governmental policy under conditions of uncertainty: A century of controversy about Saccharin in Congress and the laboratory', *Minerva* **18**: 556–74.

Stolberg, S.G. (1997), 'Expert panel rebuffs bid to absolve Saccharin', *The New York Times* (1 November): A13.

Turner, S. (2003), *Liberal Democracy 3.0: Civil Society in an Age of Experts*, London: Sage.

U.S. House of Representatives (2003), *Politics and Science in the Bush Administration*, Committee on Government Reform, Minority Staff. Washington, DC.

U.S. Senate (1977), *Food Additives: Competitive, Regulatory, and Safety Problems, Part 1*, Select Committee on Small Business, 95th Cong., 1st sess., Washington, DC: U.S. Government Printing Office.

U.S. Senate (1992), *National Institutes of Health Reauthorization Act of 1992*, Committee on Labor and Human Resources, S. Rpt. 102-263, Washington, DC: US Government Printing Office.

CHAPTER 5

MARK B. BROWN[*], JUSTUS LENTSCH[**]
AND PETER WEINGART[**]

REPRESENTATION, EXPERTISE, AND THE GERMAN PARLIAMENT: A COMPARISON OF THREE ADVISORY INSTITUTIONS

INTRODUCTION

At least since the first democracy executed its most prominent expert advisor, the relationship between democracy and expertise has been a topic of more than academic interest. Socrates was not a scientist in today's sense of the term, but like many experts today, and unlike the Sophists of his own time, he sought to make the search for truth useful to his contemporaries. The Athenians' marked lack of appreciation led Plato to the view that in a just state philosophers would need to be kings. Things have not worked out that way, but politics today has become unthinkable without the continual reliance on various forms of expertise. Expert advice enters the political process through established institutions, short-term commissions, ad hoc committees, and informal personal networks. Experts from every imaginable profession and academic discipline advise executive, legislative, and judiciary branches of government, as well as interest groups, businesses, and civic organizations of all kinds.

This chapter examines the potential contribution of expert advice to the representative tasks of the German Bundestag (Federal Parliament). We consider three advisory institutions relevant for legislative decision making in Germany, each primarily associated with one of the reference points of our analysis: enquete commissions (parliament), the Office of Technology Assessment (science), and citizen panels (the public sphere). We evaluate these institutions with respect to both the quality of their expertise and the extent of their contribution to democratic representation.

Political decision makers turn to experts for two fundamental reasons. First, they use expertise to make their decisions more reasonable, justifiable, and effective. Second, because the use of expertise gives decisions a greater claim to public acceptance, politicians hope that citizens will be more willing to accept a decision based on (or at least rationalized with) expert advice. Expertise thus serves what might be called problem-oriented and politics-oriented functions.[1] The former refers to the 'substantive' use of expertise to identify, understand, and make decisions about socio-technical problems. The latter refers to the communicative use of expertise to justify policies, as well as the strategic use of expertise to delay decisions or avoid

81

Sabine Maasen and Peter Weingart (eds.), Democratization of Expertise? Exploring Novel Forms of Scientific Advice in Political Decision-Making – Sociology of the Sciences, vol. 24, 81–100.
© Springer Science+Business Media B.V. 2009

responsibility. Using expertise to either develop or explain policy decisions allows politicians to make a justifiable claim to public acceptance. Even the strategic use of expertise might offer a justifiable way of promoting the goals for which a politician was elected, thus increasing acceptance among supporters.[2] It is important to note, however, that a justifiable claim to public acceptance cannot be equated with actual public acceptance, and the latter rarely depends entirely (and often not at all) on expertise.

Expertise thus provides only one of the resources with which politicians seek to make their decisions democratically legitimate. Legitimacy is of course a complex concept and cannot be explored here. For present purposes we want to suggest that ideally legitimate decisions require a combination of expert advice, popular involvement and acceptance, and legal authorization and accountability. Legitimacy thus has both substantive and procedural components. Neither rational and effective decisions that are publicly rejected, nor irrational and ineffective decisions that are publicly accepted, are fully legitimate.[3]

The substantive and procedural components of legitimacy roughly correspond to the two key elements of our normative conception of democratic representation: leadership and participation, sometimes conceived in terms of the 'trustee' and 'delegate' models of representation (Pitkin 1967: chap. 10). Public representatives in a democracy should neither slavishly follow nor entirely ignore public opinion. Representatives ought to promote those policies they consider to be in the public interest, and it is in the public's interest that representatives take the desires, opinions, and electoral preferences of regular citizens into account. Understood in this way, political representation does not conflict with public participation, as is often assumed, but depends on it (Plotke 1997).[4]

The relationship between technical expertise and democratic representation has long taken the form of a *scientization of politics*. Since the mid-twentieth century, expanding governmental activities and new technological risks have increased the reliance of advanced industrial states on technical advice. This has led to an expansion of the leadership component of democratic representation, usually at the expense of the participatory component. Expertise of various sorts has always played a key role in representative democracy, insofar as it helps representatives determine which policies will effectively promote the public interest (Ezrahi 1990: chap. 2). But in the context of scientized politics, experts are mistakenly portrayed as fulfilling a universal human interest in effective policy, and hence, as the public's only true representatives (Hitzler 1994: 17; Feenberg 1999: 137).

The scientization of politics has been associated with both *decisionist* and *technocratic* models of expertise (Habermas 1970: 62–80). According to the former, experts provide value-neutral information about available means, and politicians make value-based decisions about desirable ends. The legitimacy of political decisions is seen to rest not on substantive standards of rationality, nor on active public participation, but solely on the formal authorization and accountability of the decision makers. The technocratic model, in contrast, equates political legitimacy with the rationality and effectiveness of policy, replacing politics with scientific administration. Both models mistakenly assume it possible to promote the public interest without active public involvement. And both depend on an image of value-free science that has been repeatedly refuted by empirical research on the co-production of science and politics in the laboratory and public life (Jasanoff et al. 1995). Each model thus fails to fulfill the above-described conception of legitimacy: the decisionist model lacks substantive rationality, and the technocratic model lacks public acceptance and involvement.

In response to the shortcomings of scientized politics, interest groups of all kinds have sought their own sources of expertise, which when coupled with the complexity of socio-technical problems and the uncertainty of scientific knowledge, has led to a *politicization of science* – the flip-side, so to speak, of the scientization of politics (Weingart 2001: chap. 4). Paradoxically, the politicization of science has simultaneously increased expert prominence and decreased expert authority. And to the extent that experts today become associated with the interest groups that sponsor their work, the politicization of science extends interest-group representation into the realm of expertise.

A desire to restore expert authority without returning to scientized politics has led over the past thirty years to calls for the *democratization of expertise,* usually focused on efforts to expand the number and type of parties involved in technically complex political issues (e.g., Petersen 1984; Hennen 1999; Joss and Bellucci 2002). When determining research priorities, making policy recommendations, or even, less frequently, when conducting research itself, experts are increasingly expected to solicit and respond to the views of lay citizens. Efforts to democratize expertise often draw on a *pragmatist model* of expertise, according to which the values implicit in science and technology are subjected to political deliberation, and political goals are adjusted in light of the technical means available for their realization (Habermas 1970: 66).[5] Depending on the range of participants involved in such pragmatist mediation processes, commentators refer to either a *corporatist model* or a *participatory model* of expertise. The former includes representatives from government, science, and major interest groups; the latter expands the range of participants to include lay citizens (see Joss and Bellucci 2002). As long as it avoids a populist reduction of political questions to matters of subjective preference, the participatory approach more fully captures the aims of the pragmatist model than the corporatist view. Whereas the decisionist model reduces public participation to periodic elections, and the technocratic model includes no role at all for the lay public, a participatory version of the pragmatist model links the participation and leadership elements of democratic representation.

Efforts to democratize expertise have met with two distinct responses. Some see democratization efforts as nothing but a further politicization of expertise and argue instead for a return to an imagined golden age of value-free expert advice. Others claim that public participation on expert advisory committees justifies the immediate adoption of their recommendations by legislatures without further deliberation or consultation. From our perspective, each of these responses to democratized expertise lacks a coherent understanding of the relationship between expert advice and democratic representation. The first response assumes expert advisory bodies can ignore the lay public; the second asserts that by involving the public they acquire the same representative status as a popularly elected legislature. This chapter attempts to identify a conceptual and institutional space for expert advisory bodies that avoids both of these misconceptions.

Evaluating Expert Advisory Institutions

In developing criteria of evaluation, we have sought to go beyond the typical concern with the scientific validity of expertise. When expert knowledge is uncertain, controversial, and intertwined with value judgments, when many advisory commissions include non-experts, and when the political need is not so much for science but for policy relevant advice, traditional criteria of validity are insufficient. We have thus developed two criteria that combine a rough measure of scientific validity with certain aspects of the participation and leadership elements of democratic representation.

Representativeness

The criterion of representativeness refers to the degree to which advisory institutions incorporate diverse social, political, and disciplinary perspectives. With regard to scientific validity, the notion of disciplinary representativeness captures the basic idea of peer review, which typically seeks to include a wide range of perspectives from a single discipline. It is also similar to traditional scientific notions of publicity and openness to criticism, especially as they pertain to the frequent need for interdisciplinary cooperation in expert advisory processes. In politics the idea of representativeness is associated with the tradition of 'descriptive' representation, which conceives representation in terms of resemblance or similarity between representative and constituent (Pitkin 1967: chap. 4). In contrast to the 'delegate' model of representation, which employs elections or communication between elections to bind representatives to their constituents, the descriptive view assumes that descriptively similar representatives will spontaneously act as their constituents would have acted. It appears in the common expectation that representatives should possess the same demographic characteristics – race, class, gender, age, education, etc. – as the people they claim to represent. It can also be seen in the United States Federal Advisory Committee Act of 1972, which requires that advisory committees be "fairly balanced in terms of points of view represented and functions performed" (5 U.S.C. App. I, §§5(b)(2), 5(c); Jasanoff 1990: 47; Smith 1992).

It is important to note that descriptive similarity in either science or politics does not authorize representatives to act on their constituents' behalf. Nor can representatives whose claim to represent resides only in descriptive similarity be held accountable by or to their constituents, since people can be held to account only for what they have done, not for who they are (Pitkin 1967: 83–91). What descriptive representatives can do is call attention to the questions, concerns, and social perspective they share with their constituents (Young 1997; Mansbridge 1999). Evaluations of representativeness always remain contestable, however, as the relevance of any particular perspective to a particular question often becomes a controversial question itself.

A high degree of descriptive representativeness on an expert advisory committee has a number of potential benefits. First, to the extent that increasing the number of alternative perspectives on a problem improves understanding of the problem, representativeness contributes to the rationality of an advisory committee's work. The more perspectives involved, the more likely that errors and biases will be identified and corrected. Second, representativeness may increase the public acceptance of expert advisory committees, insofar as it increases the likelihood that committee recommendations will be responsive to the concerns of every social group relevant to the committee's topic.[6] Third, if the members of an expert advisory committee are publicly associated with particular social, political, or disciplinary groups, they may evoke a symbolic form of representation – that is, a feeling of being represented – among other members of those groups. Although symbolic political representation is easily misused for ideological purposes, it can also foster a sense of membership and help decrease the alienation of excluded groups from political life.

Despite these potential benefits, the representativeness of an advisory committee cannot by itself ensure the legitimacy of any decisions to which it contributes. The democratization of expertise does not alter the fact that, at a fundamental level, expertise aims at a primarily scientific rather than political form of representation. Rather than 'representing' in the sense of acting for others, expert advisory committees 'make representations of' empirical evidence, experiential perspectives, and normative claims. Expanding the membership of such committees may make their recommendations more broadly representative of available evidence and social perspectives, but it does not authorize such committees to act on the public's behalf. Similarly, descriptive representativeness may foster public acceptance of both the advisory process and any subsequent decisions, but it does not provide a measure of public acceptance.

Resonance

No matter how representative an advisory institution, if it remains ignored by policy makers and the public it will have little impact on either policy decisions or public discourse, and hence, make little contribution to democratic representation. We use the idea of parliamentary and public resonance to characterize the level of attention generated by advisory committees to their topics and activities among decision makers and the general public.

Our assessment of both representativeness and resonance focuses on the institutional design of our three selected advisory institutions. Our assessment depends in part, of course, on the actual performance of these institutions to date, especially in cases where an established pattern of activity suggests an informal institutional norm. But we are less concerned with the representativeness and resonance that these advisory institutions have achieved so far, and more with what can be expected in light of their institutional designs and their relationships with other institutions. We do not address micro-level questions regarding the quality of deliberation within these advisory institutions or between their members and the politicians they advise. Our conclusions thus take the form of hypotheses regarding the contribution to democratic representation that one might reasonably expect from each of these advisory bodies in light of the norms and incentives reflected in their institutional frameworks.

EXPERTISE AND THE GERMAN PARLIAMENT

Like the legislatures and parliaments of other advanced industrial states, the German Bundestag has over the past fifty years expanded its activities to address a wide range of technically complex problems, increasing its need for expertise. This has led to numerous changes in both the form and function of expertise (see Krevert 1993: 128f.; Thunert 2001). Expert advisory processes have become more explicitly political, more interdisciplinary, and more open to including lay citizens. Indeed, there has been a general shift away from decisionist and technocratic models and toward a pragmatist model of expertise. Many advisory bodies today, including those examined here, build explicitly on the pragmatist insight that politicians have little use for scientific knowledge as such, but rather for expertise tailored to their political needs. At the same time, however, this generally pragmatist orientation manifests itself in very different ways.

ENQUETE COMMISSIONS

The enquete commission was created as part of the Bundestag's 'small parliament reform' on 1 October 1969. Parliamentary investigative committees (*Untersuchungsausschüsse*) generally confine themselves to past instances of alleged corruption, so many parliamentarians wanted a new institution that would provide advice on emerging problems and upcoming decisions. Enquete commissions were also specifically aimed at overcoming the legislature's informational deficit with respect to the executive (Altenhof 2002: 12). According to the Bundestag's administrative regulations, any member of the legislature may request the creation of an enquete commission, and if 25 percent join the request a commission must be created. One half of the seats on every commission are given to members of the legislature, one half to invited experts. Both legislative and expert members are appointed by the parliamentary party groups (*Fraktionen*), each group receiving an allotment of seats in proportion to its number of seats in the Bundestag. Enquete commissions generally have 12–20 members. They meet periodically during a single legislative term, after which they may be reestablished by the next legislature. Commissions often prepare several interim reports, and they are required to provide a final report to the legislature at the end of each legislative term. Over twenty enquete commissions have addressed a wide variety of topics, including nuclear power, information technology, 'youth protest,' AIDS, global warming, genetic engineering, technology assessment, and the legacy of the East German dictatorship.[7]

Enquete commissions serve both problem-oriented and politics-oriented functions (see Krevert 1993: 167ff.). Insofar as they educate parliamentarians and the general public, they facilitate scientifically informed public policy and the effective resolution of political problems. They are not research institutions, however, and do not aim to resolve political conflicts by 'speaking truth to power.' Indeed, the Bundestag has repeatedly affirmed the essentially political character of enquete commissions (Altenhof 2002: 161f., 326). Beyond this general orientation toward political issues, different commissions have somewhat different purposes: some aim more to monitor and control the government, others more to seek consensus on an emerging issue.[8]

Some even appear to have primarily strategic purposes (Hoffmann-Riem 1988: 61). Whatever their purpose, because they include experts not elected by the public, administrative regulations explicitly limit enquete commissions to providing general recommendations rather than advocating specific policy measures (Altenhof 2002: 92).

Representativeness of Enquete Commissions

Given their institutional proximity to political power, one can expect the disciplinary representativeness of the expert members of enquete commissions to be lower than that of most other advisory institutions. It is no secret that the parliamentary groups select experts with the aim of garnering scientific validation for their political positions. Experts are not chosen according to their party membership, but the selection process generally involves careful screening of an expert's scientific publications in light of their political implications. Although experts occasionally surprise the party that invited them, the lines of division on an enquete commission usually run not between experts and politicians, but between the commission members from the governing coalition and those from the opposition parties, with the experts aligned with the side that invited them. Depending on the particulars of the case, this arrangement can hinder the inclusion of all relevant disciplinary perspectives (Hoffmann-Riem 1988: 63). Indeed, some expert members of enquete commissions have complained that, if they wanted to have any influence on the commission's deliberations, they had to tailor their statements to their sponsor's position (Ismayr 1996: 37).

In addition to their institutional bias against high disciplinary representativeness, enquete commissions have limited social representativeness. Although enquete commissions increasingly hold extensive public hearings, they are not legally required to involve the general public. Nor are there institutional incentives to employ demographic criteria in selecting commission members. Demographic criteria seem to play a role only when it becomes politically impossible to ignore them, as with the commission on the legacy of the East German dictatorship, which emphasized the inclusion of participants from former East Germany (Altenhof 2002: 181–3). There are no formal requirements, however, for the inclusion of traditionally excluded social perspectives, such as those of women and minorities.

A representative political composition, in contrast, is an implicit goal of the enquete commission's institutional design. Because the parliamentary members are appointed by the parliamentary party groups in proportion to the groups' electoral strength, the political makeup of every enquete commission mirrors that of the Bundestag. Insofar as the legislature is descriptively representative of the full range of political views in German society, enquete commissions will be too. The use of proportional representation in the German electoral system facilitates the representation of a relatively wide range of political views. This does not guarantee, of course, that all political views are represented, and those members of the legislature not aligned with a parliamentary party group (*Fraktionslose*) have charged that their exclusion from the appointment of members to enquete commissions reduces the commissions' representativeness (Altenhof 2002: 80–85). More generally, the direct link between the political composition of enquete commissions and that of the legislature creates an institutional limitation on complete representativeness not present in, for example, citizen panels. This lack of full political representativeness might help explain why enquete commissions have so far not addressed some of the issues that most concern German citizens, including unemployment, terrorism, German unification, security policy, and European integration (Altenhof 2002: 334f.). Nonetheless, enquete commissions probably have a higher degree of political representativeness than any other form of expert advice in Germany.

Parliamentary Resonance of Enquete Commissions

Some enquete commissions elicit far more interest from Bundestag representatives than others, due simply to the topicality of the subjects they address. Beyond the matter of parliamentary interest in their topics, the most important factor in the uptake of commission ideas and reports appears to be the efforts of the parliamentary members to mediate between the commission and the Bundestag through both informal contacts and organized workshops. Although everyone on an enquete commission has a single vote when approving the final report, parliamentary members have a certain 'home court' advantage: they are familiar with the procedures, they chair the meetings, they may bring an assistant to the meetings, and they have existing alliances and cooperative relationships with other members of the legislature (Altenhof 2002: 205). Expert members of the commission tend to have more influence in cases where parliamentarians have not yet committed themselves to a particular position on the topic. Overall, however, a commission's influence does not depend primarily on the quality or quantity of the scientific evidence assembled by the commission. Indeed, the more enquete commissions succeed in capturing the complex, interdisciplinary character of the problems they study, the more difficult it becomes to assimilate their reports to the segmented organizational structure of the legislature (Ismayr 1996: 40). Rather than scientific validity, parliamentary resonance depends on the efforts of individual parliamentarians in actively promoting a commission's work (Altenhof 2002: 203–209).

Such efforts have so far proven successful in only certain respects. With regard to the problem-oriented functions of expertise, the Bundestag has never adopted all the recommendations of an enquete commission, and no recommendations have been implemented that were not in accord with the program of the majority coalition (Altenhof 2002: 318). Enquete commissions have achieved greater parliamentary resonance with respect to the politics-oriented functions of expertise. For example, they have sometimes been effective at serving a 'pilot function': parliamentarians who are able to reach a consensus among the members of an enquete commission can expect to reach one in the Bundestag as well (Altenhof 2002: 209f.). Additionally, the most influential commission recommendations have been those that were already present in the broader public discourse (Altenhof 2002: 320). In sum, enquete commissions merit relatively high expectations for parliamentary resonance, but primarily with regard to their politics-oriented functions, and especially when combined with strong public resonance.

REPRESENTATION, EXPERTISE, AND THE GERMAN PARLIAMENT

Public Resonance of Enquete Commissions

As with parliamentary resonance, the topics of some enquete commissions simply elicit more public resonance (and mass media attention) than others. Several enquete commissions have actually been established in response to public controversies on a particular topic (Altenhof 2002: 321). In recent years, enquete commissions have sought to stimulate public interest by making their work as publicly transparent and accessible as possible (Ismayr 1996: 41). Although they remain primarily oriented toward the Bundestag, enquete commissions have increasingly made use of public hearings, symposia, and other means of involving the public in their work. The commission on the East German dictatorship, for example, heard testimony from 327 experts and concerned citizens at 24 public hearings. The commission on global climate change heard testimony from almost 500 experts (Altenhof 2002: 222–225). The commission on 'youth protest' even held one of its hearings on live television (Altenhof 2002: 322).

The interim and final reports of every commission are published by the Bundestag and occasionally by a commercial publisher as well. Commission meetings are generally not open to the public, nor are transcripts usually published, so as to spare participants public scrutiny, give them more freedom to modify their positions, and thus facilitate the search for consensus (Altenhof 2002: 209). Several commissions, however, have made their commissioned reports and other research materials available to the interested public. Some commissions have also solicited written testimony from both civic organizations and the general public (Altenhof 2002: 226). According to one assessment, enquete commissions have in recent years pursued a 'continuing dialog' with the interested public (Hampel 1991: 119). Thought it might go too far to call them "one of the most important instruments of interaction between parliament and society" (Braß 1990: 94), it seems reasonable to expect a relatively high public resonance from most enquete commissions.

THE OFFICE OF TECHNOLOGY ASSESSMENT AT THE GERMAN PARLIAMENT (TAB)

The *Büro für Technikfolgenabschätzung beim Deutschen Bundestag* or TAB was created by the Bundestag in 1990. The authorizing directive called for the establishment of an institution whose legal form, scientific competence, and interdisciplinary structure would allow it to provide advice to the legislature with a high degree of institutional independence (Deutscher Bundestag 1989). The task of establishing such an advisory body was thus appropriately delegated to the Institute for Technology Assessment and Systems Analysis (ITAS), a major research institute in Karlsruhe, Germany. The ITAS is almost entirely government funded, but it remains institutionally independent of the Bundestag. Organizational and political responsibility for the TAB is held by the Bundestag's Committee for Education, Research, and Technology Assessment, facilitated by a permanent rapporteur group, comprised of one member from each of the parliamentary party groups.

The TAB pursues a diverse program of activities aimed at, first, understanding the legal, social, and environmental potentials and risks associated with scientific and technological developments; and second, suggesting alternative options for political action, though not specific policy measures (Petermann 1994: 80; Meyer 1997: 347). The TAB does not generally perform research itself, but commissions either original research or literature reviews of existing research. Given its political independence and institutional distance from the legislature, the TAB is the most scientifically-oriented of the advisory institutions examined here. It embraces a relatively traditional, 'instrumental' conception of technology assessment, making it the most suited to the problem-oriented functions of expertise (see Peters 1996; Petermann 1999: 56).[9]

Representativeness of the TAB

Unlike enquete commissions and citizen panels, the TAB has little aspiration to either social or political representativeness. The notion of political representativeness does appear in the work of the TAB's parliamentary permanent rapporteur group (*Berichterstatter-Kreis*), a subcommittee of the Bundestag's science and technology committee. The rapporteur group has the task of turning legislators' often very general expressions of interest in research on a particular topic into concrete research proposals. The rapporteur group is supposed to remain politically neutral, with each parliamentary party group appointing only one member. Nevertheless, insofar as the work of the rapporteur group involves politically charged decisions, it may have a distinctly political influence on the topics of TAB research. This is only to say that, as an advisory institution, the TAB's work might be considered politically representative in the minimal sense that it conducts research on topics of interest to those in power. Social representativeness, in contrast, seems to play no role in the TAB's work.

With regard to disciplinary representativeness, there is little evidence that the political influence on the selection of TAB research topics extends to the research itself. Indeed, the TAB's mission is explicitly conceived as advising the entire parliament, rather than any particular parliamentary group (Beyme 1997: 160f.). Put differently, the TAB seeks to make its work representative of scientific rather than political opinion. Although the TAB staff is relatively small (currently ten scientists), a wide range of disciplines are represented, including biology, chemistry, physics, agricultural sciences, political science, sociology, and economics. Moreover, when preparing its reports the TAB commissions 5-10 external studies, seeking to solicit a wide range of scientific opinion (Grunwald 2003). It also occasionally holds interdisciplinary workshops, thus increasing the disciplinary representativeness of its projects. This generally high disciplinary representativeness is decreased somewhat by the dominance of the social and natural sciences with respect to both the TAB staff and the topics of TAB reports. Perspectives from the humanities are almost entirely absent. Given the ethical issues at the center of recent public debates on genetic research, the lack of bioethical expertise, in particular, is an important limitation of the TAB's disciplinary representativeness.

REPRESENTATION, EXPERTISE, AND THE GERMAN PARLIAMENT

Parliamentary Resonance of the TAB

The TAB's impact on legislative processes is even more indirect than that of enquete commissions, and its parliamentary resonance is not easy to assess. The parliamentary rapporteur group holds primary responsibility for ensuring that TAB reports and activities receive a hearing in the legislature. The rapporteur group attempts to bring attention to TAB advice in all of the relevant Bundestag committees. This is an enormous task for which there is rarely sufficient time and expertise (Deutscher Bundestag 2002). The reception of TAB reports is also hindered by the inevitable conflict between their interdisciplinary approach and the highly specialized character of legislative committees.

Nonetheless, TAB reports have often had an indirect effect on Bundestag decision making. They do not contain specific policy recommendations, but aim rather to provide an informational basis for parliamentary deliberation. Of the 78 reports prepared between 1991 and 2001, twenty-five were published in the official Bundestag register; nine of those resulted in Bundestag resolutions proposed by the rapporteur group (Deutscher Bundestag 2002). It appears that TAB reports contribute to the conceptualization of problems and the development of parliamentary agendas, even without being directly referenced in parliamentary debate.

Public Resonance of the TAB

Unlike the other advisory institutions examined here, the TAB generally does not seek a direct influence on public discourse on technical issues. Nonetheless, it is possible to identify a few areas in which it has achieved a certain level of public resonance. Many TAB reports are available to the public, the agency publishes a biannual newsletter, and it maintains a public website. A recent parliamentary assessment recommended that the TAB undertake more aggressive public relations work, suggesting public workshops, increased cooperation with other research institutes, and participatory technology assessment projects as ways to involve the lay public (Deutscher Bundestag 2002). Another possibility is to allow public access to the Bundestag committee meetings at which new TAB reports are initially presented. This occurred for the first time on 21 May 2003.

Finally, the TAB may be said to have a certain amount of indirect public resonance. Unlike the executive branch, the German legislature has a constitutional mandate for public transparency, which it seeks to fulfill in various ways. Plenary sessions, for example, are televised and open to the public. If TAB reports and activities find resonance within the Bundestag, they may also contribute in a roundabout way to public discussion of scientific and technological issues (see Peterman 1999: 52).

CITIZEN PANELS

Recent calls for involving lay citizens in the work of enquete commissions and the TAB pay homage to a thirty-year tradition of participatory expertise. The frequent political bias of mainstream technology assessment toward elites, as well as its epistemological bias toward technical rather than social and moral questions, has fostered a wide variety of efforts to include regular citizens in expert advisory procedures (see Saretzki 1997: 281; Joss and Bellucci 2002: 6). A number of studies have compared the various existing forms of participatory technology assessment (e.g., Rowe and Frewer 2000). We focus here on those that a) bring experts and lay citizens into dialogue with each other, b) include participants not affiliated with established interest groups, and c) address themselves to both policy makers and the general public. These criteria are most clearly fulfilled by consensus conferences, planning cells, and citizen juries, which we refer to collectively as 'citizen panels.'

Citizen panels consist of a group of 10-20 lay citizens who meet on three or four weekends to learn about and discuss a socio-technical issue, confer with an expert panel, write a report with policy recommendations, and then hold a press conference to publicize their work. Citizen panels have been sponsored by both private and governmental institutions. They aim to educate participants, stimulate public discourse, and advise decision makers on socio-technical issues. Although the precise meaning of 'lay citizen' often remains unclear, organizers expect that participants will articulate goals and values different from those of most experts and politicians. Discussions among the panelists are meant to follow a 'deliberative' model in which panelists eschew bargaining or self-interested claims in favor of reasoned argument. Even so, organizers usually allow majority and minority reports when consensus proves impossible. As of 2002 about 50 citizen panels had been organized in over fifteen countries on a wide range of socio-technical issues, including transgenic plants and animals, food irradiation, telecommunications, atomic waste, genetic testing, and stem cell research (Loka Institute 2002). In Germany, planning cells have been organized since the 1960s (Dienel 2002), and in 2001 the German Hygiene Museum in Dresden sponsored the country's first consensus conference (Schicktanz and Naumann 2003).[10]

Representativeness of Citizen Panels

Citizen panels aspire to high social representativeness, but their methods for achieving it are diverse and complex.[11] Citizen juries and consensus conferences use either a telephone poll or advertisements in local and national media to generate an initial selection pool. The organizers then draw on the pool to select a panel fulfilling a range of demographic criteria, including age, gender, education, occupation, and area of residence. Political party membership or ideology has not usually been a selection factor, suggesting that political representativeness is not a priority for most citizen panels. Planning cells rely solely on random selection to compose the panel, selecting a larger number of participants for 2-10 panels that run simultaneously (Dienel 2002: 253).[12]

Despite the widespread use of random selection in assembling citizen panels, organizers often fail to clarify whether the goal is to achieve a statistically representative sample or a demographic cross-section of the population (see Carson and Martin 2002). In a statistically representative sample, the number of people representing each significant social group is proportionate to the number of that group in the general population. Defining 'significant social group' is of course problematic, as is determining which people ought to be deemed representative of which groups (Smith

REPRESENTATION, EXPERTISE, AND THE GERMAN PARLIAMENT   93

and Wales 2000: 56–57). It is clear, however, that a panel of 10-20 members is far too small to be statistically representative of even the most relevant social groups in any of the countries where citizen panels have been organized. A cross-section, in contrast, need only have a single member from each relevant social group.[13] Nonetheless, many commentators continue to uphold the statistical sample as an implicit ideal, presumably because it seems to ensure every citizen an equal chance of participating.

There are two things to be said here. First, in comparison to other forms of citizen participation – voting, demonstrating, contacting public officials, even donating money – the number of those involved in citizen panels is extremely small. Inequalities in the probability of selection, therefore, pale in significance when compared to the enormous inequalities in most other forms of participation. Second, the use of random selection does not provide an equal *opportunity* for everyone to participate, but merely an equal *probability* of being chosen. Those chosen must accept the invitation, but those not chosen have no way to become involved. The purpose of representativeness on citizen panels, therefore, ought not to be seen in terms of its contribution to citizen participation. There is little reason, therefore, to prefer an ideal of statistical representativeness to that of a representative cross-section. Indeed, despite occasionally misleading formulations, most organizers justify their selection procedure with reference to the idea of a representative cross-section (Hörning 1999: 357; Hennen 1999: 356). This ideal standard makes it likely that citizen panels will realize the benefits of social representativeness more fully than the other advisory institutions examined here.

Unlike the lay panelists, the participants on the expert panel are not randomly selected but carefully hand-picked by organizers, usually with some degree of input from the lay panelists. The aim has generally been to achieve as much variety as possible with regard to both the fields of expertise and the range of opinion on the relevant issues (Durant 1995, 77; Joss 1995, 99–100). Like the TAB and unlike enquete commissions, citizen panels have no institutionalized incentives that would prevent a high degree of disciplinary representativeness on the expert panel. At the same time, however, most citizen panels have far fewer financial and organizational resources than the other institutions examined here. Those enquete commissions that hold public hearings with a large number of experts probably achieve a higher degree of disciplinary representativeness than a citizen panel. In terms of institutional design, however, citizen panels match the high disciplinary representativeness of the TAB.

Parliamentary Resonance of Citizen Panels

Most citizen panels seek to impact legislative decision making in some way. This goal is most obvious in those countries, such as Denmark and the Netherlands, where citizen panels are institutionally linked to the national legislature (see Gloede and Hennen 2002). In a recent survey, Danish legislators said that by lessening their dependence on biased experts and uninformed citizens, consensus conferences had made important contributions to legislative decision making (Grundahl 1995: 38; Joss 2000: 347–48). There is also evidence that the Danish Parliament's decisions to ban food irradiation and to prohibit companies from demanding DNA-profiles of

94 MARK B. BROWN, JUSTUS LENTSCH AND PETER WEINGART

their employees were influenced by consensus conferences on those topics (Andersen and Jaeger 1999: 335). The nature and degree of this influence, however, is very difficult to assess.

Unlike enquete commissions and the TAB, the parliamentary resonance of citizen panels may conflict with their goal of offering a critical perspective on legislative decisions and stimulating public debate. Indeed, overemphasizing the potential for political influence might create incentives for panelists to tailor their recommendations to the exigencies of legislative decision making. It appears reasonable, then, to expect less direct legislative resonance from citizen panels than from the other advisory institutions examined here.

Public Resonance of Citizen Panels

Beyond their potential impact on policy makers, most citizen panels seek to influence both the general public and the panelists themselves. Such influence might take the form of changes in people's substantive knowledge of the topic of the panel, their procedural knowledge about the policy process, or their reflexive knowledge of themselves as citizens (Guston 1999: 469f.). With regard to influence on the panelists themselves, most participants report having learned a lot about the topic of the panel, and many claim to have an increased interest in science and technology policy well after the conclusion of the panel. Most seem to take the task very seriously, and they appreciate being taken seriously as political actors (Smith and Wales 2000: 60f.). Although there is little to be said against such educational effects, one might ask whether they are an effective way of improving citizen involvement in science and technology policy. Whatever educational and empowering effects citizen panels have on participants, their possibilities in this regard pale in comparison to those of traditional civic organizations, political parties, and interest groups – all of which, however, might well benefit by adopting the pragmatist approach to expert advice evident in citizen panels.

Given these considerations, it seems that the more important potential of citizen panels lies in their impact on the general public. Given sufficient media coverage, citizen panels can serve as crystallization points for public discussion of sociotechnical issues. As one commentator puts it, a citizen panel "should act as a two-way link between public debate and the representative decision-making institutions. As such, it draws on, and seeks to represent, public discourse on science and technology, as well as advancing it by feeding the results of the assessment procedure back into it" (Joss 1998: 21). The degree of media coverage and public interest has been very different for different citizen panels, but their institutional design equips them to speak at least as well to the general public as to political decision makers.

CONCLUSION

The social, political, and disciplinary representativeness of each advisory institution examined here depends primarily on its procedures for selecting participants and its use of external resources. A comparison of enquete commissions and citizen panels shows that the latter aim for a higher level of social and disciplinary representativeness. Citizen panels have not tended to emphasize political representativeness, but there is nothing to prevent them from doing so, and it is an implicit aspect of their

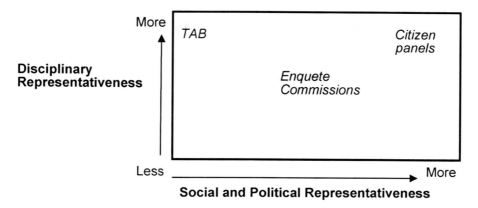

Diagram 1: Social, political, and disciplinary representativeness in the design of three advisory institutions

announced goal of generating the most inclusive deliberation possible. Although the superior resources of enquete commissions, and especially their use of public hearings, may often lead to a level of representativeness higher than most citizen panels, their reliance on an openly partisan procedure for selecting participants sets a lower institutional standard for representativeness. Whereas the organizers of citizen panels may be asked by those excluded from a particular panel to justify the exclusion with reference to the panel's topic, the parliamentary party groups that organize enquete commissions have an electoral mandate to pursue their party program and may well be justified in tailoring their selection of participants to this purpose. The TAB aspires to high disciplinary representativeness, but it has thus far sought neither political nor social representativeness (see Diagram 1).

With regard to parliamentary and public resonance, our analysis suggests that both are usually very indirect. Each of the three advisory institutions aims primarily to shape the identification, understanding, and discussion of socio-technical problems, rather than the resolution of those problems. The TAB and enquete commissions both aim for a high degree of parliamentary resonance. Enquete commissions have in recent years increasingly defined their task in terms of public resonance as well. Citizen panels are the most dependent on public resonance, relying on mass media coverage to influence both public discourse and legislative decision making. Whereas citizen panels seek parliamentary resonance via public resonance, the TAB has the potential of reaching the public through its impact on legislative debate. The TAB has thus far sought little direct resonance with the general public (see Diagram 2).

It would be a mistake to assume that each of these advisory institutions should seek to maximize its resonance and representativeness on all levels. Not only would this exceed the resources of most institutions, it might in some cases be counterpro-

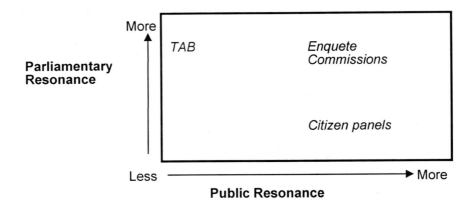

Diagram 2: Parliamentary and public resonance in the design of three advisory institutions

ductive. Citizen panels, as noted above, might compromise their ability to stimulate critical public debate by seeking too much parliamentary resonance. The TAB might sacrifice some of its disciplinary representativeness if it sought to also include a wide range of social and political perspectives. Indeed, the differences among these advisory institutions allow each to make a distinct contribution to the legislature's task of democratic representation. The TAB provides politically relevant but otherwise neutral scientific expertise; enquete commissions facilitate pragmatist negotiation over competing policy options in light of expert testimony; and citizen panels provide an informed but potentially critical perspective from outside the framework of mainstream scientific and political institutions.

The contribution of each advisory institution to democratic representation might also be seen in terms of political leadership and participation, as mentioned above. Contributions to leadership appear primarily in advisory committee reports; contributions to participation can be seen in processes of public consultation. The TAB thus fosters the Bundestag's efforts to exercise political leadership when legislators draw on its work. Legislators advised by the TAB are better able to devise and rationally justify effective public policies, which stimulates public confidence that the legislature is acting responsibly in the public interest. The greater the TAB's disciplinary representativeness and parliamentary resonance, the greater its contribution in these respects. The TAB as currently structured, however, has little potential to contribute to the participation element of democratic representation.

Enquete commissions, in contrast, are well suited to contribute to both leadership and participation. They promote leadership by introducing expertise into legislative decision making; they foster participation by providing a platform for the articulation of citizen interests, to which legislators can respond. Due to their unique institutional location, enquete commissions might well benefit by maximizing all the forms of representativeness and resonance examined here.

Given that the recommendations expressed by citizen panels have been refined and informed through expert advice and collective deliberation, they can make a limited contribution to parliamentarians' efforts to provide democratic leadership. When legislators seek to identify the public interest, as opposed to expressed citizen opinions, they are justified in paying special attention to the recommendations of citizen panels. But citizen panels are not authorized to act on the public's behalf, and they should not be treated as mini-parliaments or the authentic voice of the people. Insofar as they stimulate public debate and help parliamentarians learn about citizen concerns, citizen panels are suited to fostering the participation element of democratic representation. Like the other institutions examined here, citizen panels should be wary of sacrificing their contribution to one aspect of democratic represenation for the sake of another.

*California State University, Sacramento, USA
**Universität Bielefeld, Germany

ACKNOWLEDGEMENT

This paper builds on research funded in 2003 by the German Parliament's Office of Technology Assessment, the *Büro für Technikfolgenabschätzung beim Deutschen Bundestag* (TAB). For their suggestions and assistance, we would like to thank the TAB, as well as colleagues at Bielefeld University's Institute for Science and Technology Studies, and participants at the preparatory workshop for this volume, held in December 2003 at Basel University.

NOTES

[1] Bimber (1996: 36) draws a similar distinction between the 'analytical' and 'rhetorical' uses of expertise.

[2] For example, a politician might use the recent completion of an expert study as an excuse for requiring a decision before the opposition can assemble supporters.

[3] Note that public acceptance should ideally be expressed through both informal political communication (e.g., public discourses, civic organizations) and formal procedures of popular and governmental decision making (e.g., popular elections and voting in parliament). Note also that if public rejection of governmental decisions takes the form of voting those responsible out of office, this does not by itself lessen the legitimacy of the preceding decisions. But without the authorization and accountability that elections provide, decisions cannot be democratically legitimate in the fullest sense. This view of legitimacy seeks to combine the Weberian emphasis on legal procedures, the focus in empirical political science on public preferences, and the normative concern with rational justification prevalent in recent theories of deliberative democracy (see Connolly 1984).

[4] This view of political representation as an ongoing process of interaction between state institutions and civil society is broadly compatible with recent interest in processes of 'governance.' The state is cast in the role of facilitating the resolution of public problems through cooperative networks, rather than 'engineering' or 'steering' society from the top down.

[5] Note that the pragmatist model does not imply an elimination of the boundaries between science and politics (Weingart 2001: 159). Science and politics each maintain their own rationalities as distinct social systems, each structured around a different set of rules, norms, incentives, and goals. The preserva-

tion of boundaries between science and politics makes efforts to mediate across them all the more important (Guston 2000).

[6] Such effects are not guaranteed, of course, and in some cases deliberation among diverse participants may exacerbate rather than ameliorate disagreements (Warren 1996).

[7] Sixty-eight percent of the 'experts' sitting on enquete commissions are associated with a scientific or scholarly institution; 16 percent have no such affiliation and can be considered interest group representatives (Altenhof 2002: 183f.).

[8] Whether a problem- or politics-orientation prevails may have a lot to do with which party holds the majority of seats on the commission. Those commissions led by the Social Democratic Party (SPD) have tended to see their task as stimulating public discussion, those led by the Christian Democratic Party (CDU) have focused on proposing legislative solutions (Altenhof 2002: 167). Given that opposition parties have a strategic interest in ongoing discussion, while governing parties have an interest in policy solutions, this difference in style may be traceable to the fact that the SPD was in the opposition during most of the enquete commissions held to date, i.e., from 1982 to 1998.

[9] Several studies have examined the scientific quality and political influence of TAB activities (e.g., Petermann 1994; Peters 1996; Meyer 1997; Paschen 2000). They devote little attention, however, to questions of democratic representation or political legitimacy (see Grunwald 2003). It is also worth noting that most existing studies on the TAB have been performed by TAB staff members. Similarly, a recent evaluation of the TAB by the Committee for Education, Research, and Technology Assessment was conducted in close cooperation with TAB staff (Deutscher Bundestag 2002).

[10] In addition to the members of the panel itself, citizen panels rely on a small organizing committee to select participants and manage the overall process. Many citizen panels also have an independent steering committee to provide advice, and they often employ a professional facilitator to ensure the fairness and efficiency of their deliberations. Some citizen panels even make use of secretarial services to assist with preparing the final report.

[11] The following discussion of representativeness on citizen panels is developed more fully in Brown (2004).

[12] In contrast to the handpicking of participants by the organizers, random selection lends the process a sense of objectivity. It may thus increase the panel's contribution to both the rationality and public acceptance of subsequent decisions. Additionally, since random selection conveys the notion that anyone may have been invited to participate, it may give the general public a sense of being symbolically represented by the panel (Renn et al. 1995: 353).

[13] Some have argued that minority groups should actually have disproportionately more members, to ensure that their perspectives on the issue receive a fair hearing (Mansbridge 1999).

REFERENCES

Altenhof, R. (2002), *Die Enquete-Kommissionen des Deutschen Bundestages*, Wiesbaden: Westdeutscher Verlag.

Andersen, I.-E. and B. Jaeger (1999), 'Scenario workshops and consensus conferences: Towards more democratic decision-making', *Science and Public Policy* **26**, 5: 331–40.

Beyme, K. (1997), *Der Gesetzgeber: Der Bundestag als Entscheidungszentrum*, Opladen: Westdeutscher Verlag.

Bimber, B.A. (1996), *The Politics of Expertise in Congress: The Rise and Fall of the Office of Technology Assessment*, Albany, NY: State University of New York Press.

Braß, H. (1990), 'Enquete-Kommissionen im Spannungsfeld von Politik, Wissenschaft und Öffentlichkeit', in T. Petermann (ed.), *Das wohlberatene Parlament: Orte und Prozesse der Politikberatung beim Deutschen Bundestag*, Berlin: Edition Sigma: pp. 65–95.

Brown, M.B. (2004), 'Citizen panels and the concept of representation', paper presented at the annual meeting of the Western Political Science Association, Portland, Oregon, March 12.

REPRESENTATION, EXPERTISE, AND THE GERMAN PARLIAMENT 99

Carson, L. and B. Martin (2002), 'Random selection of citizens for technological decision making', *Science and Public Policy* **29**, 2: 105–13.

Connolly, W. (ed.) (1984), *Legitimacy and the State*, New York: New York University Press.

Deutscher Bundestag (1989), *Technikfolgenabschätzung und -bewertung beim Deutschen Bundestag*, BT-Drs 11/4749.

Deutscher Bundestag (2002), *Technikfolgenabschätzung (TA): Beratungskapazität Technikfolgenabschätzung beim Deutschen Bundestag – ein Erfahrungsbericht*, BT-Drs 14/9919.

Dienel, P.C. (2002), *Die Planungszelle: Der Bürger als Chance*, 5th Ed., Wiesbaden: Westdeutscher Verlag.

Durant, J. (1995), 'An experiment in democracy', in S. Joss and J. Durant (eds.), *Public Participation in Science: The Role of Consensus Conferences in Europe*, London: Science Museum, pp.75–80.

Ezrahi, Y. (1990), *The Descent of Icarus: Science and the Transformation of Contemporary Democracy*, Cambridge, MA: Harvard University Press.

Feenberg, A. (1999), *Questioning Technology*, London: Routledge

Gloede, F. and L. Hennen (2002), 'Germany: A difference that makes a difference?', in S. Joss and S. Bellucci (eds.), *Participatory Technology Assessment: European Perspectives*, London: Centre for the Study of Democracy, pp. 92–107.

Grundahl, J. (1995), 'The Danish consensus conference model', in S. Joss and J. Durant (eds.), *Public Participation in Science: The Role of Consensus Conferences in Europe*, London: Science Museum, pp. 31–49.

Grunwald, A. (2003), 'Technology assessment at the German *Bundestag*: "Expertising" democracy for "democratising" expertise', *Science and Public Policy* **30**, 3: 193–8.

Guston, D.H. (1999), 'Evaluating the First U.S. Consensus Conference: The impact of the citizens' panel on telecommunications and the future of democracy', *Science, Technology, & Human Values* **24**, 4: 451–82.

Guston, D.H. (2000), *Between Politics and Science: Assuring the Integrity and Productivity of Research*, Cambridge, UK: Cambridge University Press.

Habermas, J. (1970), *Toward a Rational Society: Student Protest, Science, and Politics*, trans. J.J. Shapiro, Boston: Beacon Press.

Hampel, F. (1991), 'Politikberatung in der Bundesrepublik: Überlegungen am Beispiel von Enquete-Kommissionen', *Zeitschrift für Parlamentsfragen* **1**: 111–33.

Hennen, L. (1999), 'Partizipation und Technikfolgenabschätzung', in S. Bröchler, G. Simonis and K. Sundermann (eds.), *Handbuch Technikfolgenabschätzung*, Vol. 2, Berlin: Edition Sigma, pp. 565–71.

Hitzler, R. (1994), 'Wissen und Wesen der Experten. Ein Annäherungsversuch – zur Einleitung', in R. Hitzler, A. Honer and C. Maeder (eds.), *Expertenwissen: Die institutionalisierte Kompetenz zur Konstruktion von Wirklichkeit*, Opladen: Westdeutscher Verlag, pp. 13–30.

Hoffmann-Riem, W. (1988), 'Schleichwege zur Nichtentscheidung: Fallanalyse zum Scheitern der Enquete-Kommission "Neue Informations- und Kommunikationstechniken"', *Politische Vierteljahresschrift* **29**, 1: 58–84.

Hörning, G. (1999), 'Citizens' panels as a form of deliberative technology assessment', *Science and Public Policy* **26**, 5: 351–9.

Ismayr, W. (1996), 'Enquete-Kommissionen des Deutschen Bundestages', *Aus Politik und Zeitgeschichte* **27**: 29–31.

Jasanoff, S. (1990), *The Fifth Branch: Scientific Advisors as Policymakers*, Cambridge: Harvard University Press.

Jasanoff, S., G. Markle, J.C. Petersen and T. Pinch (eds.) (1995), *Handbook of Science and Technology Studies*, Thousand Oaks, CA: Sage Publications.

Joss, S. (1995), 'Evaluating consensus conferences: Necessity or luxury?', in S. Joss and J. Durant (eds.), *Public Participation in Science: The Role of Consensus Conferences in Europe*, London: Science Museum, pp. 89–108.

Joss, S. (1998), 'Danish consensus conferences as a model of participatory technology assessment: An impact study of consensus conferences on Danish parliament and Danish public debate', *Science and Public Policy* **25**, 1: 2–22.

Joss, S. (2000), 'Participation in parliamentary technology assessment: From theory to practice', in N.J. Vig and H. Paschen (eds.), *Parliaments and Technology: The Development of Technology Assessment in Europe*, Albany: State University of New York Press, pp. 325–62.

Joss, S. and S. Bellucci (eds.) (2002), *Participatory Technology Assessment: European Perspectives*, London: Centre for the Study of Democracy.

100   MARK B. BROWN, JUSTUS LENTSCH AND PETER WEINGART

Krevert, P. (1993), *Funktionswandel der wissenschaftlichen Politikberatung in der Bundesrepublik Deutschland: Entwicklungslinien, Probleme und Perspektiven im Kooperationsfeld von Politik, Wissenschaft und Öffentlichkeit*. Hamburg and Münster: LIT Verlag.

Loka Institute (2002), 'Danish-style, citizen-based deliberative "consensus conferences" on science & technology policy worldwide', at http://www.loka.org/pages/ worldpanels.htm, accessed on 23 February 2002.

Mansbridge, J. (1999), 'Should blacks represent blacks and women represent women? A contingent 'Yes', *Journal of Politics* **61**, 3: 628–57.

Meyer, R. (1997), 'Das Büro für Technikfolgenabschätzung beim Deutschen Bundestag', in R.G. von Westphalen (ed.), *Technikfolgenabschätzung*, 2nd. ed., München and Wien: Oldenbourg, pp. 340–65.

Paschen, H. (2000), 'The technology assessment bureau at the German parliament', in N.J. Vig and H. Paschen (eds.), *Parliaments and Technology: The Development of Technology Assessment in Europe*, Albany: State University of New York Press, pp. 93–124.

Petermann, T. (1994), 'Das Büro für Technikfolgenabschätzung beim Deutschen Bundestag', in A. Murswieck (ed.), *Regieren und Politikberatung*, Opladen: Leske + Budrich, pp. 79–99.

Petermann, T. (1999), 'Technology assessment units in the European parliamentary systems', in N.J. Vig and H. Paschen (eds.), *Parliaments and Technology: The Development of Technology Assessment in Europe*, Albany: State University of New York Press, pp. 37–61.

Peters, J. (1996), *Technikfolgenabschätzung und politische Verantwortung des Parlaments: Das Büro für Technikfolgenabschätzung beim Deutschen Bundestag in einer politikwissenschaftlichen Wirkungsanalyse*, Bremen: Univ., Forschungszentrum Arbeit und Technik.

Petersen, J.C. (ed.) (1984), *Citizen Participation in Science Policy*, Amherst: University of Massachusetts Press.

Pitkin, H.F. (1967), *The Concept of Representation*, Berkeley: University of California Press.

Plotke, D. (1997), 'Representation is democracy', *Constellations* **4**, 1: 19–34.

Renn, O., T. Webler and P. Wiedemann (1995), 'The pursuit of fair and competent citizen participation', in O. Renn, T. Webler and P. Wiedemann (eds.), *Fairness and Competence in Citizen Participation: Evaluating Models for Environmental Discourse*, Dordrecht: Kluwer Academic Publishers, pp. 339–67.

Rowe, G. and L.J. Frewer (2000), 'Public participation methods: A framework for evaluation', *Science, Technology, & Human Values* **25**, 1: 3–29.

Saretzki, U. (1997), 'Demokratisierung von Expertise? Zur politischen Dynamik der Wissensgesellschaft', in A. Klein and R. Schmalz-Bruns (eds.), *Politische Beteiligung und Bürgerengagement in Deutschland*, Baden-Baden: Nomos, pp. 277–313.

Schicktanz, S. and J. Naumann (eds.) (2003), *Bürgerkonferenz: Streitfall Gendiagnostik. Ein Modellprojekt der Bürgerbeteiligung am bioethischen Diskurs*, Opladen: Leske + Budrich.

Smith, B.L.R. (1992), *The Advisers: Scientists in the Policy Process*, Washington, DC: Brookings Institution.

Smith, G. and C. Wales (2000), 'Citizens' juries and deliberative democracy', *Political Studies* **48**: 51–65.

Thunert, M. (2001), 'Politikberatung in der Bundesrepublik Deutschland seit 1949', in U. Willems (ed.), *Demokratie und Politik in der Bundesrepublik 1949–1999*, Opladen : Leske + Budrich, pp. 223–42.

Warren, M. (1996), 'What should we expect from more democracy? Radically democratic responses to politics', *Political Theory* **24**, 2: 241–70.

Weingart, P. (2001), *Die Stunde der Wahrheit? Zum Verhältnis der Wissenschaft zu Politik, Wirtschaft und Medien in der Wissensgesellschaft*, Weilerswist: Velbrück.

Young, I.M. (1997), 'Deferring group representation', in I. Shapiro and W. Kymlicka (eds.), *Ethnicity and GroupRights*, NewYork: NewYork University Press, pp. 349–76.

CHAPTER 6

STEPHEN TURNER

EXPERTISE AND POLITICAL RESPONSIBILITY: THE *COLUMBIA* SHUTTLE CATASTROPHE

One of the major conflicts between the principles of democratic politics and the practical reality of expertise in public decision making takes place in connection with responsibility. The basic principles of democracy include some notion of political responsibility, usually understood to take a bi-directional form, in which the relation between ruler and ruled takes the form of representation and the relation in which the ruled control the ruler takes the form of accountability. The means by which the people assure that the persons who politically represent them are responsible to them vary among democratic regimes. Even within the general framework of representative liberal democracies the character of political responsibility varies considerably, both within a particular regime and between political regimes. Moreover, typically a modern state employs many devices apart from simple parliamentary representation, and very often these devices, such as juries, commissions, independent judges, law lords, and the like predate (or are based on models that predate) parliamentary government. Yet they involve their own notions of responsibility that resolve, ultimately, into a form of political responsibility.

Expert opinion has traditionally played an ambiguous role in basic constitutional arrangements, falling neither into the ruler nor the people side of this structure. Mill, in his classic constitutional text *Representative Government* (1861), was exceptional in that he was careful to discuss what he called council as a part of representative government. But he was vague about how it relates to responsibility, and for good reason. There are some very striking differences between the kind of responsibility that can be assigned to decision makers and the kind that can be assigned to experts, either in making decisions which are the products of expertise or expert opinion or simply expressing expert opinion in the context of a process leading to decisions. In this chapter I will explore these ambiguities in terms of a case study: the *Columbia* Shuttle catastrophe of 2003.

RESPONSIBILITY: THE *COLUMBIA* SHUTTLE

In its sharpest form, the contrast between the personal character of decision-making, which leads to the notion of personal political responsibility, and the liability for the decisions made,

101

Sabine Maasen and Peter Weingart (eds.), Democratization of Expertise? Exploring Novel Forms of Scientific Advice in Political Decision-Making – Sociology of the Sciences, vol. 24, 101–121.
© Springer Science+Business Media B.V. 2009

is represented in Max Weber's famous colloquy with Erich Ludendorff, in which the people are given the ultimate power to judge the actions of their leader on their behalf and to send him to the gallows if he fails.

> Max Weber: In a democracy the people choose a leader whom they trust. Then the chosen man says, "Now shut your mouths and obey me. The people and the parties are no longer free to interfere in the leader's business."
> Ludendorff: "I could like such a democracy!"
> Max Weber: "Later the people can sit in judgment. If the leader has made mistakes, to the gallows with him" ... (Marianne Weber 1975: 653).

Expert opinion, by definition, can never be judged in this way. Characteristically, expert opinion is not even formulated in the first person. Nor are individuals thought to be personally liable for opinions competently formulated within the limits of professional standards of expert knowledge. Put more simply, expert knowledge is impersonal knowledge expressed or formulated by individuals, while political responsibility is personal and judged by collective processes.

The reality beyond these abstractions, however, is more complex. The individuals who express technical opinions do so as persons, and the expression itself is governed by a kind of ethic. Speaking for science has a form that resembles representation, though the representation is of the science community and its opinions (see Turner 2002). But a personal element remains. The somewhat paradoxical notion of unbiased technical opinion, which is to say a technical opinion that is genuinely impersonal and therefore genuinely purely technical, is exemplary of the problem: if an opinion were purely technical, it is difficult to see what sort of question of bias could arise. The term itself implies that technical opinions are inherently impure, a mix of the personal and the technical, and that one can be held responsible for technical errors and biases, but not for one's unbiased opinions (see Thorpe 2002). Unbiasing ones *expressions* of opinion, however, serves to obscure or eliminate one's personal responsibility for them, and thus shifts responsibility to decision-makers.

When something goes wrong in a situation that involves matters that are primarily technical and includes political responsibility, these abstract paradoxes about the relation between technical expertise and responsibility turn into real administrative and political issues, and bureaucratic distinctions between council or expert opinion and participation in decision-making break down. The *Columbia* shuttle catastrophe of 2003 is an example of this transformation and concretization of the problem of responsibility. The catastrophe itself immediately led to the creation of a classic pre-democratic council mechanism for delegating technical questions to experts, namely a commission. The commission was given a task which drew on a specific and highly developed political tradition of boards of inquiry designed to assign responsibility for particular outcomes in complex events. What is striking about the *Columbia* case, which embodies the paradoxes of political impersonality, is that technical advice itself appears to have been responsible for the decision, so that the task of the board was to evaluate expert opinion and decision-making made on the basis of it. I will suggest that the phenomenon of expert knowledge here, as elsewhere, presents pervasive challenges to the notions of political responsibilities that underlie democratic theory, which can be highlighted by asking the Schmittian questions of "Who decides?" and "Who exercises unquestionable discretion?" (Schmitt 1985: 32–3). The answer here appears to be that no one can be assigned responsibility – an anomalous result for democratic theory.

In general, the production of expert knowledge is controlled through the indirect means of professional certification and recognition, the concentration and control of opinion through methods of peer review and evaluation that typically do not directly assess the truth or validity of expert claims but only their minimal professional adequacy, but which nevertheless, through a series of complex redundant and overlapping mechanisms, generate consistent expert conclusions that fall within a small range. Thus responsibility for expert opinion is diffused by the very mechanisms which create the phenomenon itself, though as we will see there are some important exceptions to this. For the most part, however, the political power experts exercise over one another is limited, episodic, and indirect. Journal editors, board certifiers, and degree granters exercise discretion, but their powers are indirect: they do not determine truth, but evaluate competence, admissibility of claims, and the application of standards. This is the personal substructure of the impersonality of expertise. The wide distribution, indirectness, and uncontroversial character of personal decision-making allow for the illusion of impersonality. Political decision-making, in contrast, involves direct mechanisms of command and control as well as direct mechanisms of punishment for failure that enforce responsibility. The impersonality of expert claims, in contrast, removes the possibility of assessing responsibility, and even expert error itself becomes something that only experts themselves are in a position to assess. One intriguing aspect of the *Columbia* case is that the commission was used precisely for this purpose: as a means of using expert opinion to evaluate the actions of an expert bureaucracy, NASA.

The *Columbia* case provides a quite stark example of the transformation and evanescence of the concepts of representation and accountability in the face of expertise. In the immediate aftermath of the catastrophe, in numerous newspaper editorials and statements by congressmen there was a call for blame to be assessed and the blameworthy punished. For the public, holding individuals responsible was an essential aim, and their representatives took this as part of their own political responsibility. The expressions of the desire to hold individuals responsible were typified by those of Senator Ernest Hollings, who observed that this was "an avoidable accident" (a condition of responsibility), and noted that "in similar circumstances a Navy ship captain would be 'cashiered'" (*St. Petersburg Times*, September 4, 2003).

In response to the widespread public desire for accountability, the administration of NASA, in anticipation of an external investigation, immediately appointed an internal review committee which was assigned the task of investigation. The committee was promptly dismissed in the press as an inside job, in spite of the care that had been taken to include persons who were not directly involved in the decision-making. An additional board of outsiders and former NASA officials was appointed under the title of the *Columbia Accident Investigation Board* (CAIB) and began to work (*Orlando Sentinel*, February 13, 2003). The CAIB chose to mimic aircraft accident investigation boards. In this case, 200 investigators were put to work producing a report which cost nearly a half billion dollars.

The CAIB was advised not only by technical experts but also by top line management theorists and social studies of science scholars. It came to conclusions contrary to the conclusions of interested outsiders and citizens as well as Congressmen, many of whom rejected implicitly the CAIB account in favor of one that emphasized responsibility. The Board and the managers responsible for the shuttle also took opposed views of the problem of responsibility. The differences between them point to fundamental issues over the aggregation of expert opinions and over the assignment of responsibility for errors and incompetence that bear directly on the fundamentals of liberal democratic theory. In what follows, I will show why these are intrinsic to the problem of expertise.

Ending the Problem of Responsibility: NASA and the CAIB

Among the first acts of the commission was to choose a model for their inquiry, namely aircraft accident boards, that enabled them to avoid the task of assigning personal responsibility. The ostensible reason for this was that it would be impossible to expect candid testimony to a commission which was inquisitorial and judicial at the same time. But the choice also reflected generic and important difficulties with applying judicial and quasi-judicial concepts to expert inquiries. One problem was this: the Board was compelled to rely for its judgements not only on the testimony of the potentially guilty, which would be an ordinary situation in criminal cases, but on their claims as experts – in which bias would inevitably be an issue. Although the reasoning behind particular decisions, the differences of opinion about these technical decisions, and so forth, was more or less accessible to the members of the commission, who functioned as near-experts, and was accessible to the advisers to the commission as even nearer experts, neither the commission nor its advisers could reasonably hope to develop a completely independent understanding of these issues on their own. So the testimony was a curious combination of exculpation and expert witnessing of a kind that no ordinary court would have permitted as testimony.

Public discussion of the inquiry continued to use the phrase "smoking gun," which is to say the language of criminal accountability, and, despite the Board's self-limitations, the reality that the findings of the Board would be translated into some sort of punitive action was never far from either the public discussion or the thinking of the participants in the process, both on the Board and among the NASA employees. Because technical causes implied prior administrative choices, there was good reason to believe that some sort of personal accountability would result.

Public speculation about causes, and particularly about damage produced by the large chunks of foam that were seen to have been shed from the fuel tanks at liftoff, began immediately after the event. The release of internal NASA e-mails involving a frantic discussion among engineers of the possible consequences of foam shedding added to the speculation. The e-mails included a characterization of a potentially catastrophic scenario which proved to be an eerily accurate prediction. The puzzle then became the question of why nothing was done, even after a plausible professional opinion within the NASA team had made this prediction.

The initial reaction of the NASA bureaucracy to foam speculation was to attack its credibility. The NASA director Sean O'Keefe, a manager with science studies management background rather than a physicist, disparaged what he called "foamologists" and characterized the effect of the foam on the *Columbia* as similar to the effect of a styrofoam container on hitting the front of a pickup truck. To call the adherents to the foam explanation "foamologists" implied that there was no form of technical expertise at its base – foamology is a parodic name for a nonexistent technical discipline. Presumably these comments reflected what O'Keefe had been told by senior technical managers. As it turned out, he was relying on technical opinions that turned out to be grossly in error.

The magnitude of the error became apparent in the middle of the inquiry when an experiment was conducted by the CAIB simulating the kind of high speed foam hits that the orbiter had actually been subject to. It was clear that the foam was more than sufficient to do the kind of damage that would have doomed the orbiter – and that the foamologists were right. But there was a crucial ambiguity about the results that bore on responsibility: the discussions within NASA during the flight itself involved the heat tiles that protected the orbiter during re-entry, which was assumed to be the weakest link, while the actual damage, as the CAIB reconstructed the evidence, was to the leading edge of the wing, made of reinforced carbon-carbon (RCC) (CAIB/NASA 2/18/2004; see Sec. 5: 1–60; Sec. 12: 1)

THE SHUTTLE MANAGER'S DEFENSE

In the middle of the investigation, after it had begun to be clear that the technical cause of the catastrophe was foam, the mission managers, those responsible for the decision about what to do about possible foam damage, held an extraordinary press conference to get their side of the story into the media. This side of the story was part of the evolving NASA response to the events which is itself revealing as a confirmation of the claims about the nontechnical causes of the event that form a large part of the CAIB report. In this interview the NASA managers centrally involved in the decision-making retreated to a second line of defense, which for our purposes poses the critical issue because it so neatly encapsulates the core problem of expertise in relation to fundamental concepts of responsibility and representation.

The actions that came into question were taken by Linda Ham, mission manager, her deputy, and their senior technical advisor, in response to an informal request by the debris team for satellite imagery of the bottom of the orbiter to see if it was possible to identify holes or damage to the tiles of the kind that would have resulted from a foam strike. This request was made shortly after the launch itself, after a frantic e-mail discussion within the team about the videotape of the launch, where extensive foam shedding had been observed. There is some question about the exact nature of the discussions between the various advisors concerned to have an image of the tiles and the flight management team, but the response of the management team was extremely interesting. The management team not only declined to entertain their request, but declined to do so on procedural grounds, the ground that it had not come through proper channels, or to put it in the peculiar bureaucratic language of Ham herself:

> We have read news reports that the mission management team had declined a request for outside assistance and if you read through the transcripts, you'll note that the mission management team never addressed a request for outside assistance because it never came up in any of the meetings. It never came up to me personally (Harwood 2003).

There was some confusion over what amounted to a proper channel for such a request, but the request was discussed at length not only with Ham's assistant but with her technical advisor, and Ham herself intervened and "killed the request," according to reports (*Orlando Sentinel*, July 1st, 2003) because the "possible request" had failed to come through proper channels, and thus was not a bureaucratically genuine request, or in the language of NASA, a "requirement" (Langewiesche 2003: 81).[1] But to track down the source of the request, as she put it:

> I began to research who was asking, and what I wanted to do was find out who that person was and what exactly they wanted to look at, so that we could get the proper people from the ops team together with this people or group of people, sit down and make sure that when we made the request, we really knew what we were trying to get out of it (Harwood 2003).

This statement was widely interpreted as indicating that she was trying to identify the source of the request in order to punish and intimidate the requester, a desire that becomes more relevant in the context of some other remarks by Ham about the model of decision-making for the process.

Ham, in the interview, argued explicitly against holding anyone responsible, on the grounds that "we do operate and we communicate, and everything that we do, we do it as a team," commenting that "if the system fell down, we will fix the system, but it is really difficult for me to attribute blame to individual personalities or people" (Harwood 2003). Ham had good reason to attempt to deflect responsibility, as she had made the critical problematic decision. In her capacity as manager, she had decided that the problem was something that the team could not and would not have done anything about until after the mission anyway. But she did so with the public assent of relevant technical officers. Formally, what she said was true: This was a collective decision, though a collective decision made through a process that, as she herself observed, might have been flawed.

Part of the problem of Ham's responsibility arose through issues of intent. She appeared to have been motivated by managerial considerations (that in retrospect seemed trivial and inappropriate) about the cost of getting the images and about the inconvenience and cost to other elements of the mission of maneuvering the orbiter into a position such that photographs could be taken. Moreover, Ham raised questions about the adequacy of the paperwork and rationale for dismissing foam events on previous flights, and indeed, because there had been such a large number of these events, the problem was one of which she was at least minimally aware. But here again her motivation appeared to be simply to construct an adequate rationale for dismissing the event which could be based on the precedent of previous rationales for dismissing it; 'appeared to be' because if she had taken the possibility of damage seriously she would have opened a full scale inquiry while the flight was still going on. Her concern with the paperwork on previous flights seems to have included no interest in or willingness to actually do anything with this paperwork, such as reading it or having it re-examined for its relevance. And her bureaucratic approach to the problem of foam shedding seemed, in retrospect, also to be inappropriate and callous. Her interest in citing it appeared to be to use it as bureaucratic authority for her own actions. Her comment that she hoped the paperwork was good, in the context of her failure to address the question of whether it was, indicated that she had no interest in the question, despite the fact that she had to be aware, as her comments indicate, that these rationales were often not especially good.

Because it would have at least been possible to consider strategies to save the orbiter at this point, this decision would have been in a similar context, such as military context, culpable, and intent would have been relevant. But she employed an Eichmann-like defense of 'following procedures.' And the top administrators of NASA seemed not only willing but eager to treat this as not only acceptable but competent managerial action. As O'Keefe later said, she was an excellent manager and other divisions of NASA would be eagerly competing for her managerial services. In the military context, she would likely be court-martialed, as Senator Hollings observed, though in those contexts as well the reliance on procedure would have been at least partly exculpatory.

Why did this apparently strange acceptance of the 'procedural' defense seem normal within NASA? One answer is this: NASA was tracking 5,396 individual hazards that could cause major system failures in flight, and about 4,222 of those could either threaten mission success or cause the loss of the crew. Waivers had been granted on 3,223 of those known problems (Halvorsen 2003). Thus, there were many possibly catastrophic unsolved problems with the shuttle with similar paperwork histories. Foam shedding was a case in which the paperwork was extensive, the problem had been experienced repeatedly before, progress had been made in dealing with it, and in no case had the damage been severe enough to cause the loss of the orbiter. So there was a substantial rational probabilistic ground for thinking that the problem would not arise in the case of this foam strike either, and Ham herself noted in the key meeting that "the flight rationale [for going ahead with launching after the foam strike in October was] that the material properties and density of the foam wouldn't do any damage," and suggested looking at the data from a 1997 flight where there had been debris damage (Sawyer 2003). In the face of 'paperwork' of this kind, the reliance on procedure becomes less puzzling. No manager could hope to assess 5,000 hazards on their merits: that was the task of engineering teams that had formal responsibility to speak up, through channels, if they believed the issue needed to be addressed. Ham relied on those procedures because there was no practical alternative to doing so.

The second line of defense opened by Ham pointed to some reasons why this would have been a defensible action purely on grounds of the NASA model of handling expert opinion. As one NASA official was quoted, "she did the best she could do given the information she had. She talked to people she trusted, she listened to the analysis" (Harwood 2003). This in itself is a peculiarly ambiguous formulation, but it is nevertheless, from the point of view of responsibility, an interesting one.

As it happens, the technical advice which seems to have decisively sealed the fate of the requests for imagery was from Dan McCormack, a senior structural engineer, who told the team that the analysis of possible tile damage by the contractually responsible Boeing team showed no serious threat, and that the RCC wing edge might show some coating damage, but no "issue" (Sawyer 2003). At this meeting another engineer spoke up to agree with him. The Houston-based NASA engineer, Alan Rodney Rocha, who had expressed strong concern, backed down after this analysis.

What should we make of these errors? In a model of decision-making in which there is a well-developed division of labor such that particular technical decisions and particular competencies match perfectly, there is no problem in assigning responsibility. If a particular cable breaks, the person whose job it is to make decisions about the cable and to be expert about everything relating to the cable is responsible. But a complex system typically cannot be understood in this way and new systems contain novel unknowns that cannot be understood in this way.

The technological division of labor (and expert knowledge) that works for automobiles or aircraft carriers, which are well-understood technologies, will not work for a poorly understood technology because the relations between the parts that correspond to the division of labor may be different than expected. In this case at a certain very simple level there was a part with a problem tile, and therefore a corresponding body of 'expertise.' This primary level of expertise did not fail: no one was proved wrong with respect to anything that they properly gave a technical opinion on. The tiles, as far as the CAIB could tell, were *not* the cause of the catastrophe (James et al. 2004: 9). The error was in taking this legitimate piece of expertise and interpreting it to have significance it could not bear. This is a problem of aggregating technical opinion, or deciding what to make of it.

The NASA method for aggregating expert opinion involved two key elements: to rely on an elaborated division of labor involving teams responsible for specific engineering systems used in the orbiter and the launch system, as well as teams for specific problems, such as debris. Each of these teams reported to higher level mission teams which operated in a similar way: they had a managerial hierarchy, with persons who were formally responsible, but operated, with respect to technical issues and to a large extent also with respect to managerial issues, as a team which required consensus. Technical disagreements were taken seriously, and a strong rule of procedure was to insist on data-based discussion. Both of these elements of procedure worked against minority opinion, a point to which we will return, but also created a situation which made the zone of relevance of particular kinds of expertise ambiguous, so that a person participating in a consensus decision who was not genuinely expert in some area could overreach and deliver a strong message that could affect the consensus.

Among other things that need to be said about consensus in the context of large bureaucracies is that the consensus that occurs in the face of unequal, and indeed hierarchical relations between those participating in decisions, is not necessarily 'representative' of expertise in the same sense as consensus that emerges among equals. Even in the case of a consensus among formal equals, it is typically necessary to reach agreement through the artificial means of specifying what counts as agreement, such as through voting, because a genuine consensus is so difficult to actually achieve. In the case of hierarchical bureaucracy, 'consensus' is easier to achieve because it is typically only consent, not overwhelming agreement between independent experts. People on the bottom tend to go along with strong judgements made at the top out of fear for their careers; there is a strong selective bias in any bureaucratic career structure that punishes those who reject prevalent ideas.

But even in the case of genuine consensuses of independent experts, there is a question of domains of expert competence. And this is a conundrum that is central to the problem of the political significance of expert knowledge. Ham was caught in the following paradox of authority. She was not herself an expert on the issue in question, and was compelled to rely on expert judgement. But because she was not an expert, she could not independently determine who was genuinely expert. To put this in another way, there is a paradox of managerial omniscience: if the manager is sufficiently omniscient to determine who is in fact expert, they will be virtually as expert as the expert herself, and indeed, this would be the limiting case; so the closer one approaches to the limiting case, the less there is any need for expertise, counsel, and expert advice in the first place. Not surprisingly, she relied on 'the system.' In this case she relied on senior technical advisors, who screened and evaluated technical advice and comment from engineering teams, and on data-driven presentations at meetings in which quite large numbers of managers and engineers, representing different teams, participated.

AGGREGATING EXPERT OPINION: THE ROLE OF META-EXPERTISE

Some method of aggregating expert opinion was necessary, simply as a result of the complexity of the system. But a subtle and important change in the character of the expertise in question occurs as the technical issues move up to the highest level of decision-making, and a related change occurs in the meaning of consensus. If we ask who was responsible for the fact that the team went wrong in arriving at a consensus, the greatest responsibility falls on the people who influenced the consensus. The culpable figure in the story now becomes the expert who overreached his competence with bad but persuasive advice. One could imagine constructing from this an appropriate, enforceable ethic of expert opinion that made this kind of error culpable. But this cannot be a workable procedure, as it would have the effect of chilling discussion, which should consist in sharing information and mutual persuasion. Ham was accused, perhaps correctly, of creating an atmosphere in which engineers were intimidated and afraid to speak out. But Ham relied on the fact that the relevant technical advisors had a formal responsibility to raise concerns. It was Ham's zeal to hold them responsible for raising questions that chilled the atmosphere and prevented discussion. So a method of limiting this kind of discussion by making people responsible for the consequences of presenting an opinion to a group would have a predictable effect of restricting the content of the discussion further.

It is evident that assigning responsibility for *expressions* of technical opinion is a peculiarly difficult matter.[2] The engineers who invented the scenario which correctly predicted the course of events once the orbiter was damaged made a great point to the newspapers that their predictions were only speculative, absolving themselves from responsibility for not pushing their arguments harder, which was their clear formal responsibility, on the solid grounds that their scenario *was* only speculative, that is, not a technical opinion in the engineering sense, grounded on known principles and data. One of the key figures, Bob Daugherty, who did the damage assessment that indicated "what would happen to the shuttle's left tires if extra heat got inside the landing gear department because tiles were damaged" and sent a series of e-mails about this, indicating his frustration with the lack of response, was later to insist that "his messages have been grossly misinterpreted and were neither warnings nor concerns, but "just engineers talking," an interpretation that the investigators rejected.

The distinction between "just talking" and something else is nevertheless interesting, for the something else would have to be 'taking managerial responsibility.' The 'team' structure of NASA's consensus system for employing expert opinion requires engineers to take this dual role, and thus, in effect, to police themselves with respect to the expression of their opinions, which has the effect of requiring them to distinguish between offstage technical speculation and opinions for which they are accountable as members of a management team. One might say that it was this part of the 'system' that failed, because a whole series of engineers who expressed doubts in the course of 'just talking' failed to take the next step of invoking formal procedures that would have made those doubts into managerial actions (*Orlando Sentinel*, March 23, 2003).

Their actions were thus defensible: they lacked confidence in their own predictions. So, however, were Ham's. Ham pointed to some reasons why this would have been a defensible action purely on grounds of the NASA model of handling expert opinion. Ham argued that she relied on the best technical advice available. The advice that she received clearly was not correct, nor was it in fact the advice that the most appropriate and competent experts even in NASA itself would have given her. Nevertheless, in another significant sense, it was the best advice. It represented the assessment made by her own senior technical advisors of the technical advice given by the debris team and the advice given by other technical specialists, through a formal process involving responsibility, however burdensome, in the case of the lower level engineers, those responsibilities were. Only in retrospect could she have known that this heavily vetted advice was wrong.

What these various 'bests' in the category of advice indicate is what is perhaps the fundamental problem in understanding technical advice in terms of these familiar bureaucratic and democratic concepts: accountability is difficult. The potential role of an aggressive advocate of a technical opinion raises some other interesting questions. Terms like 'consensus' and 'team' have a formal meaning as well as an informal reality. The formal reality is that particular group dynamics, group-think, submission to a dominant or stubborn minority, a general reluctance to make waves, and the sense of safety of apparent majority opinion, may be the operative determinants of outcomes. The formalistic idea that expertise can be pooled through discussion ignores these processes. It also conceals an important change in the nature of the expertise that is supposed to be collectively produced. We might call the expertise of the competent practitioner operating in an understood domain of practice 'primary' expertise. What these bodies are designed to produce, however, is expertise about expertise: meta-expertise. The errors of the *Columbia* management team were failures to be expert about expertise – errors in judging the relevance, probative character, and implications of 'primary' expertise claims about tile damage and about debris damage that arguably was literally correct. The Boeing analysis, for example, included caveats that had to be ignored or judged irrelevant to make the report useable in the decision process. With this we enter into the interesting zone of ambiguity.

The Ethics of Speaking as an Expert

The effect of these considerations is not to eliminate notions of responsibility so much as shift them to experts functioning as councillors. But the problem of 'responsibility' for council is a difficult one, especially since the distinction between council and representation itself seems to imply that the responsibility falls on the recipient of the council who holds actual decision-making responsibility. Nevertheless, with respect to the concepts of scientific expertise and engineering expertise the traditions contain very strong notions of responsible expression of opinion, and yet another, even more revealing model, is to be found in medical expertise. In the case of scientific expertise, it takes the form of an expectation to observe the distinction between what is known to science, that is, what is accepted as knowledge that is genuinely scientific by at least a virtual or presumptive consensus of scientists, and 'speculation' or opinion. Scientists learn to speak in both ways and to distinguish the two and routinely criticize one another for crossing the line between the two. In engineering, the tradition is to draw a distinction between that which can be fully and predictably calculated according to known formulas and which results in predictable outcomes, and that which is not. One of the interesting examples of drawing this line occurs in connection with the concept of 'software engineering.' The claim has been made that software engineering is a misnomer because the sheer complexities of software result in bugs and problems that cannot be fully predicted or reduced to formula, and therefore cannot be, in the full sense of word, 'engineered.' Medical expertise is governed by a different set of imperatives in which 'speaking for' the medical consensus is not as central, because a balance needs to be struck between the need to respond, even in the absence of complete knowledge, to suffering and the Hippocratic admonition of 'do no harm.' This leads toward a greater emphasis on responsible speculation. Some years ago the chemist Linus Pauling suggested that vitamin C might be a response to cancer. Within the limits of chemists' notions of responsibility for utterances, he could reasonably say that this was a plausible hypothesis. But from the point of view of medical expertise, it was insufficiently plausible to justify the potential suffering it could cause for those who, predictably, over-optimistically took this advice and failed to avail themselves of more effective treatments, and was, therefore, denounced as irresponsible. Yet physicians themselves are commonly willing to try treatments that are not sanctioned by any medical consensus if they work for the individual or appear to work in other empirical cases even though the underlying mechanisms are not known. For many drugs, for example, most of the prescriptions written are 'off-label,' meaning they are written for conditions other than those they were tested to treat. But this is justified by the physician's own assessments, for which she can be held responsible, of the potential benefits and risks of the uses, which typically are based on experience – 'empirical' in medical parlance – rather than a full understanding of the relevant mechanisms and disease processes.

This more conservative ethic bears directly on the *Challenger* and *Columbia* cases, because each situation involved the assessment of data which either could not be, or had not yet been, reduced to the level at which it could be 'engineered,' but which nevertheless could justify a certain judgement about the relevance of the facts and their importance and epistemic weight. The problem with *Challenger* was a seal on a joint that was known from prior post flight examination to have behaved in an anomalous way, but which had also successfully functioned under exactly the stresses that the joint was being engineered to function under. Here, understanding and empirical knowledge did not match, just as empirical knowledge and medical science often do not match. In the face of uncertainties of this kind, it is common to employ heuristics based on the 'empirical' successes of the past. The fact that some theoretical consideration had failed to accurately predict a problem in the past was a reasonable ground for ignoring it, or if not for completely ignoring it, giving it a smaller weight in decision-making. In the case of *Challenger*, some engineers *were* concerned about the problem, but in the end a specific kind of consensus process within a managerial system, which we will discuss in the next section, overrode those concerns, and even those who voiced the concerns went along with the consensus. It is this process that many commentators pointed to as eerily reminiscent of the *Columbia* catastrophe and indeed it involved the same kinds of discrepancies between data and actual past experience: foam had not, despite hitting the orbiter 57 times, done sufficient damage to threaten a flight (*Florida Today*, August 26, 2003).

In the case of *Columbia*, there was an additional problem involving the data and its presentation that resulted from the way in which NASA routinely responded to this kind of problem. As I have mentioned, NASA discourse operated on a principle of accepting only data-based claims, yet was reluctant to spend the money and effort required to collect the relevant data when, empirically, things were working. The difficulty with data thus took on the following structure. A concerned engineer or engineering group would seek authorization to conduct the relevant engineering study, have this request denied, and then be told that its concerns were not backed by data and therefore could not be considered. Typically this response came from two sides of the communication system. On the management side, decisions were made to prioritize the relevant requests for research; on the expert advice side, claims made by lower level engineers were routinely dismissed for not being sufficiently data driven. Not surprisingly, in this information-poor environment, it was difficult to persuade either consensus groups or higher-ups of positions involving any concern, and the overriding desire of groups to be seen to be problem-free discouraged arriving at a consensus that would have placed them in conflict with announced goals of both mission management and the agency itself. This meant that, in addition to the usual mechanisms which served to punish employees for failing to act in accordance with consensus, there was a mechanism that had the effect of threatening to punish the groups who formed the consensuses themselves.

Should experts restrict their claims to those matters about which they are truly expert? There is a complex issue here. At the close of the World War II, James Bryant Conant, reflecting on his experiences, was very direct about his sense that experts, particularly scientists, routinely had hobby horses; this implied that they routinely overreached their technical competence in advancing their own views. He also recognized that the consensuses of scientists were themselves often in error, even with respect to their most fundamental beliefs. It was this aspect of Conant's thought that is amplified in his protégé Thomas Kuhn's *Structure of Scientific Revolutions* ([1962] 1996), where conformity to the paradigm was taken to be characteristic of normal science. Conant's suspicion of the overreaching expert is particularly relevant here. Who is the best judge of the limits of expertise? The question seems to lead us to exactly the problem of expert omniscience with which we ended our discussion of managerial responsibility.

If experts are genuinely the best judge of the limits of their own expertise, it would seem that expertise in effect had no limits, or to put it differently, that they were not only experts in the primary sense but meta-experts, at least in relation to everything involving their primary expertise. They would have to have the kind of birds-eye view of the domains of expertise that their own expertise related to that permitted them to say what they were expert about, and what they were not. In practice, experts often must apply the tools of their own expertise to situations that are outside of the core of their expertise, and for the scientist, this is simply the means of producing scientific results and scientific progress. For the engineer, it may be the necessary element in engineering creativity, which obviously cannot consist merely in the application of known formulas to previously solved problems, but may involve the discovery of new applications that can be then reduced to formula, or the making of novel applications. In the case of the expert who dismissed the foam problem and those experts who were above him and took his opinion as their own, backed by the 'consensus' of experts at the lower level, one wonders whether this was merely, so to speak, a 'normal error' of expertise rather than something genuinely culpable. Without the kind of expert knowledge possessed by others, would this expert have known that his particular extension of his knowledge was simply erroneous? Without what we might call omniscience about expertise, omniscience about what the limits of expertise are in particular cases of expert knowledge, we would not be able to make these judgements. And there is no reason to think that this particular expert possessed this particular kind of second order omniscience.

There is also a problem relating to the use of consensus and the conflict between a refined division of labor and the pooling of expertise to produce a consensus. If the point of consensus is not merely to diffuse responsibility and produce what I have elsewhere called fact-surrogates (Turner 2003: 41–3) for managers to operate with, but is to actually facilitate and improve expertise or improve consensuses which are in some sense better than the limited expert knowledge within particular expert domains in the division of expert labor, there must be some discursive process that allows for the dialectical evolution of this improved consensus. And whatever else the discursive ethic of this form of discussion might be thought to consist in, it does seem that the familiar considerations of Mill's *On Liberty* ([1859] 1975), of collective fallibilism and willingness to tolerate dissent and indeed even to encourage it, must be part of the ethic.

The problem, however, is that this particular model, and even its elaborated and modernized version as promoted by Habermas under the title of 'the ideal-speech situation,' in which second-order considerations about the assumptions and interests of the parties involved is made open to discussion in order facilitate the rationality of the discussion, is undermined and fundamentally transformed by the asymmetries of knowledge that are intrinsic to the situation of expert knowledge itself. The participants are not equal and all the relevant issues are not fully accessible to all the participants. Typically they are fully accessible to *none* of the participants. As an expert one participates in these discussions with a command of a particular limited range of techniques of analysis, experience, tacit knowledge, practical understanding, and so forth, and the considerations which one can bring to bear on the questions to which the consensus is addressed are necessarily limited to those considerations which one not only understands best but understands the grounds for best. In the setting of discussions with other experts with different domains of expertise, it is impossible to completely eliminate these asymmetries through discussion – Habermas' model. Discussions, however prolonged, will not produce the same tacit knowledge, practical experience, theoretical understanding, and so forth in all of the participants in the discussion. So the discourse must be of a kind that relies on trust in the expert claims of others, and also in at least some of their meta-expert claims. And this means that participants must rely on, or at least make judgements of, the self-discipline of the other participants – particularly of their observance of the distinction between expertise proper and meta-expertise, questions involving the limits, relevance, and character of the participants' own expertise.

This means that the kinds of considerations that motivated Mill, namely that free and open discussion would allow the formation of a consensus in which the participants understood one another cannot fully occur. There is an ineliminable role for meta-expertise: judgements about the competence of others to make the claims they make, and this adds an ineliminable limit to any given participant's capacity to inquire into the grounds for claims made by others. Put differently, the condition of omniscience about expertise – perfect meta-expertise – cannot obtain. Necessarily then, this means that the notion of responsibility for assent to a consensus is no longer readily applicable. One necessarily assents to things that one does not fully understand. But one cannot be held responsible for that which is beyond one's powers. One can scarcely be said to be responsible for an outcome about which one necessarily relies on others and on one's inevitably limited non-omniscient judgement of their capacity to make those judgments. But since every participant in the dialog is in this situation, in effect no one is responsible. Moreover, any stringent ethic of self-limitation in discussion would defeat the purpose of open discussion entirely and simply amount to delegating the decision or elements of the decision to those people who claimed to have relevant expertise or were assumed by the division of labor of expertise to possess that expertise. This suggests that we really have no usable model of consensus here and that these consensus processes are a kind of black box, which we can relate to as consumers in terms of their reliability, but not in any reasonably full sense understand, much less assign individual responsibility for.

NASA, MANAGERIAL RESPONSIBILITY, AND CONSENSUS

I have introduced this rather odd concept of meta-expertise, expertise about expertise, not because I think there could be expertise about expertise, but because there would need to be such a thing in order for many of the common claims about experts to make sense. It was an interesting feature of the *Columbia* inquiry that there was a significant amount of speculation, for example web pages, about the competence of the NASA managers, especially Linda Ham. The suggestion was made that anyone with elementary knowledge of physics would have realized that foam could have caused significant damage, and it was implied that her lack of this knowledge explained her mistaken actions. A more sophisticated variant of this argument was made by Hugh Pennington, who argued that

> when the known weakness in the design of the solid rocket boosters was discussed at the *Challenger* pre-launch conference, one senior manager was unhappy. He was told to "take off his engineering hat and put on his management hat." He did, and the launch proceeded – to catastrophe. With *Columbia*, the team examining the effects of the insulating foam that had peeled off the enormous external fuel tank and hit the shuttle at 500mph (creating a hole in the wing that led to the craft's destruction on re-entry) made numerous requests for imagery to be obtained to check for damage. Managers were not interested, such strikes had happened many times before and were classified as not posing a critical threat. Imagery requests were denied. When asked why the team had not pressed their requests harder, the engineers "opined that by raising contrary points of view about shuttle mission safety, they would be singled out for possible ridicule by their peers and managers" (Pennington 2003).

Put into practical organizational terms, this reasoning would suggest that there should be a complete separation between the expression of expert opinion and decision-making that no expert *should* be forced to put on a "management hat." It further suggests that sanctions for expressions of opinion – even informal sanctions such as "possible ridicule" – should be forbidden. This gives us a theoretical model for the aggregating of expert opinion for decision-making: experts express opinions for which they are not accountable, even informally, to their peers; decision-makers decide what to do with these opinions, and they are the only participants to have any responsibility. Needless to say, this is a piece of utopianism.

Having the expert put on the managerial hat makes them accountable for their opinions. It explicitly makes them accountable not merely for their technical opinions, but for something more – weighing the importance, evidential base, risks, and consequences of having their opinion taken seriously, which is to say it holds them responsible for meta-expertise about their own expertise. Is this also utopian? One view might be that the answer to the question of who is a better meta-expert about a given area of technical knowledge than the expert in this area is 'no one,' and that ordinarily, experts should be managed by experts of the same kind, and define their own scope of decision-making competence as well as their expertise. This was certainly the position of the atomic scientists in the West during the two decades after the Bomb. They believed, because they were technical experts about nuclear weapons, that they were uniquely and even solely qualified to make judgements about nuclear disarmament, about which weapons programs should be pursued, and about nuclear strategy. Eventually, however, a vast array of other kinds of experts staked a claim to expertise on these topics, and a body of informal meta-expertise guided both decision-making and the use and assessment of expert claims.

The atomic scientists, however, *sought* managerial power and responsibility, and complained about the inability of the decision-makers to understand them or respect their claim to special expertise with respect to these meta-expert topics. This model *is* another piece of utopianism, at least when it conflates the distinctions between technical knowledge and what I have been calling meta-expertise of how to assess the claims of experts in decision-making. It implies that experts should simply be delegated managerial authority in those domains about which they are expert, and that their own claims about the significance and relevance of their expertise should be accepted, even where the political consequences, about which they are not expert, are enormously significant.

The NASA system was an attempt to deal with expertise in a practical way by holding those who expressed opinions responsible for their opinions, as they bore upon decisions. "Just engineers talking" was not discouraged, and obviously, in this case, occurred. But there clearly was a problem of aggregating opinions in public discussion, which was shown in the many comments that participants in the decision-making process made about the risks of expressing concerns. There was a gap in perceptions between the underlings and the top mission managers about the quality of the atmosphere for discussion. From the point of view of the underlings, there was a sense that anything that was raised with managers had to be sugar-coated, that messengers with bad news would be punished, and so forth. The managers, in contrast, placed faith in the formal procedures, which obliged many people to raise these concerns if, in their professional opinion, the concerns were valid. And they also placed faith in reliance on hard data while failing to provide adequately for the collection of relevant data. And they seem to have been exceptionally blind to the ways in which pressures to perform and to conform with the consensus not only imposed responsibility for expressions of opinions but effectively prevented opinions from being aired. The CAIB investigators interviewing the mission managers identified this as a serious failing. In the course of the CAIB inquiry, Ham was asked

> "As a manager, how do you seek out dissenting opinions?" According to him [the investigator], she answered, "Well when I hear about them . . ." He interrupted. "Linda, by their very nature you may not hear about them." "Well, when somebody comes forward and tells me about them." "But Linda, what techniques do you use to *get* them?" She had no answer (Langewiesche 2003: 82).

One answer she could have given was that it was the formal responsibility of her subordinates to raise these questions. She chose not to blame others, however, and appealed to the idea that, because the mission workers were a team, no one should be held responsible. But the failure to 'speak to the opposition' and otherwise maintain an appropriately open kind of discussion points, as the CAIB investigator intuited, to a deeper issue with meta-expertise. Part of the point of the discursive processes of a large technical project of this kind, without the kind of stable division of knowledge spheres characteristic of well-developed technologies, such as the automobile, is to provide a collective surrogate for the necessary meta-expertise. Discussion, sharing perspectives and concerns, is the means of constructing the meta-expertise necessary for decision-making, but also, in a situation in which experts are responsible for meta-expertise about their own areas, provides the expanded understanding necessary for participants to grasp how their own domains of responsibility relate to that of others.

As important as this domain is, it is also, necessarily, the least accountable of all. The point of discussion is to arrive at a meta-expert climate of opinion. But the content of this climate of opinion itself is no one's responsibility, nor is the content a matter of expertise proper. At most, issues of *process*, such as those the CAIB investigator raised with Ham, can be treated as matters of responsibility. So the inclusion of managerial responsibility in this fashion has an ironic consequence. It makes the manager-councillors more circumspect about their claims, but at the same time frees them from responsibility for outcomes. The outcomes are the products of consensus for which no one is formally responsible.

CONCLUSION

What happened to the call to hold NASA officials responsible? There are four answers to this, and a coda which suggests that all of the discussions were misguided. The first is the answer given by the CAIB, which, in addition to its discussion of the technical causes, blamed the NASA 'culture.' The second is the response of NASA employees and former employees, which was curiously mixed. A number of different persons were identified as culpable, but perhaps the strongest reaction was that Linda Ham was a scapegoat. The third is the response of the politicians, who eventually gave up on the question of responsibility. The fourth is that of NASA management, which professed acceptance of the conclusions of the CAIB, but did what was in effect, the opposite. The coda is this: the newest understanding of the cause of the foam shedding and the foam hit itself suggests that this particular foam-shedding event was significantly outside any past experience.

One of the oddities of the CAIB report and the administration of NASA is that the board relied very heavily on social science knowledge, including the work of Karl Weick, the revered organizational social psychologist and former editor of *Administrative Science Quarterly*, and Diane Vaughan, the author of an influential book on the *Challenger* (1996). The report appealed, as Vaughan did in her book, to the organizational sociologist Charles Perrow's classic book on normal accidents (1984). The advisors to the board included Harry Lambright, who was a longtime professor in the Syracuse Science and Technology Administration program. Sean O'Keefe, the head of NASA, was both a student and a professor in the same program. So the participants were steeped not only in social science lore but in the STS tradition as it pertained to these kinds of decision-making processes.

What these participants shared was a commitment to a body of organizational behavior theory that itself served to shift issues of responsibility from individual managers to managerial processes and structures. The origins of modern managerial thinking in 'scientific management' had been based on an attack on 'the pressure system' and argued for the theory that there was a 'one best way' for performing tasks that workers themselves could not discover and hence should not be held accountable for. The Human Relations approach that followed shifted responsibility for such things as workers' feelings to managers, who came to be thought of almost as therapists. So the appeal to an organizational level of analysis already implies the diminution of the notion of responsibility and perhaps even its relegation to the medieval torture chambers of premodern organization practice.

When the CAIB dealt with the problem of responsibility, it did two things. It pointed its finger at fundamental problems in the way in which NASA was managed that were not problems that could be assigned to any single past or present administrator or administrative action. The issues were matters of process, and then 'culture,' a more recent organizational behavior concept. The notion of culture served as a useful stand-in for the guilty party, but since cultures cannot be held responsible, it had the effect of eliminating the notion of responsibility entirely.

So it is perhaps useful to close with a brief discussion of the culture argument and its implications and validity. One of the peculiarities of Vaughan's book on the *Challenger* was that the culture concept was applied to a body of decision-making practices that on another approach would have simply seemed rational. Engineers working with advanced and complex technologies, especially with technologies which are not mass produced, are routinely faced with the problem of understanding how parts fail while operating with relatively limited data. The best analogy here, perhaps, is the work of the engineers and mechanics of race cars. This is a technology in which one routinely tries to get more out of the machinery than other people have, and thus is always straining at the limits of the machinery, which routinely breaks. The failures provide data for strengthening, fixing, and redesigning parts. Safety is of course a concern, but technical decision-making relies, necessarily, on heuristics applied to actual experience. Every decision Vaughan attributed to 'culture' in the book would just as plausibly be attributed to the reliance on these heuristics,[3] and, as I have argued here, reliance on meta-expert judgements is ineliminable. The use of the heuristics or pragmatic justifications is only a part of meta-expertise, but, in any complex system with uncertainties, it is a necessary part.

The 'culture' explanation had practical implications, and here the NASA response becomes relevant. On the one hand, O'Keefe took public responsibility for the task of changing the culture. On the other, he rejected the means of changing the culture that best fit the situation. Some background here is relevant. In the early 1990s, when culture became a managerial concept, there was a spectacular and largely successful attempt to change the manufacturing culture at Ford Motor Company, in order to get managers to emphasize quality over productivity, as manufacturing quality had been a long-standing Ford problem. Some managers, particularly the head of a major New Jersey manufacturing plant, resisted or ignored the newly imposed 'culture.' As an important part of the strategy of cultural change, he was fired, and the reasons for his firing were widely circulated. Public execution, in short, is a major device for culture change. The other devices, such as charismatic 'change-agent' leadership, are more problematic, and less relevant to the NASA situation. O'Keefe, however, insisted that there was no need for a "public execution" (Cabbage and Shaw 2003), and did precisely the opposite of what Ford did: he reassigned staff members implicated in the report, allowed others to retire, but praised others, including Ham, and never circulated or even publically acknowledged the reasons for these personnel actions. Even the CAIB members accepted this. As one of them put it, "Do you want their heads on a fence someplace?" and added, "Rather than listen to what he says, watch what he does" (Leusner et al. 2004).

The predictable result of this, as Aneil Mishra, a Wake Forest professor who studies organizational change put it, would be this:

> People will read between the lines and make up their own stories and sentences and paragraphs. And they will either be wrong, distorted or they may be right. He went on to say, Sean O'Keefe needs to be telling people why those 14 or 15 people were replaced, who replaced them and why. He predicted that "if he doesn't start doing that, the culture will change, but it will be for the worse" (Leusner et al. 2004).

Barry Render, a NASA consultant and business school professor, made a similar point.

The message [from O'Keefe] is that the old system is still in place. Clearly, someone had to take blame, but they just got a lateral transfer [or] were due for retirement anyway, so that's not a big deal (Leusner et al. 2004).

When Admiral Harold Gehman, the head of the CAIB, responded to Senators pressing him on the question of accountability, he had said that, "if someone – the administrator of NASA or the head of this committee, wants to find out whose performance was not up to standard, it's all in the report." O'Keefe himself said there would be "no ambiguity on the question of accountability at all" (Cabbage and Shaw 2003). But NASA employed a routine organizational ritual that deflected the problem of responsibility further. As Robert Hotz, who had been on the *Challenger* review board put it, "you hang it on the procedures and the organization. The manager is automatically removed" (Leusner 2003). Because this would have happened anyway, as a sacrificial propitiation, it meant nothing.

Eventually a large body of opinion formed in support of the idea, as Barry Render put it, "There isn't one person to blame" (Leusner 2003). Ultimately even the politicians, who were most adamant about holding individuals responsible, such as Representative Dana Rohrabacher of California, a member of the Congressional supervising committee, in the end were persuaded that this was extremely difficult, perhaps impossible, to do. Those who persisted found the 'culture' analysis of the CAIB to be an obstacle. "... I'm trying to get past this 'culture' finding and fix responsibility," Senator Hollings said (Pianin 2003). But he could not.

The debate about responsibility continues on-line as I write this, a year after the event. The discussion is strikingly complex and inconclusive. A significant body of NASA opinion has concluded that Ham was a scapegoat, but also accepted that formal responsibility had to be taken by decision-makers to satisfy the demand to blame and punish someone, however ritualistically. A new finding by NASA explains why this foam incident was in fact unlike previous foam incidents, in which the foam had "shed" or peeled off in small pieces and fallen down along the tank and largely missed the shuttle. This time, according to the new analysis, a suitcase-sized chunk, propelled with explosive force by liquified air that collected beneath the foam, struck the orbiter, apparently causing a large gash in the wing. If this analysis is correct, the event was genuinely anomalous, and even more firmly beyond the assignment of responsibility.

University of South Florida, USA

NOTES

[1] I omit any discussion here of the communications problems that resulted from those back channel dealings in which Ham's response may have been misunderstood by lower level engineers to mean that the issue had been taken care of. This is discussed in detail in Langewiesche (2003: 81–2).

[2] A parallel case is presented by Charles Thorpe (2003: 539–46) in a discussion of the culpability of J. Robert Oppenheimer for defective advice with respect to the hydrogen bomb, which he was suspected

of having opposed for political reasons. This long-discussed, never resolved, case shows the difficulty of assigning responsibility for technical opinion.

[3] Phil Engelauf, the formally responsible senior engineer, explained the response to the foam problem in this way: "We've had incidences of foam coming off the tank throughout the history of the program and the same management processes that I think got us comfortable that that was not really a safety of flight issue have been allowed to continue, rightly or wrongly." The problem had been analyzed by Boeing, but, as he put it, "we got the wrong answer on the analysis"(Harwood 2003). The analysis, of course, came with many caveats and was not literally "wrong". The error was a meta-expert error, taking the available facts to be sufficient reason to ignore the foam problem.

REFERENCES

Cabbage, M. and G. Shaw (2003), '*Columbia* investigation: Senators take turns confronting NASA', *Orlando Sentinel,* September 4, 2003.

CAIB/ NASA (7/08/2003), *Accident Investigation Team Working Scenario,* Sec. 12, p. 1; see Sec. 5, pp.1–60, <http://www.caib.us/news/working_scenario/default.html> access date: 2/18/2004.

Columbia Accident Investigation Board (2003), 'Foam hitting orbiter nothing new', *Columbia Accident Investigation Board Report, Florida Today* (2003, August 26), Vol. I–VI, http://www.caib.us (access date: February 2004).

Halvorsen, T. (2003), 'Better analysis needed for critical flaws', *Florida Today,* August 26, 2003.

Harwood, W. (2003), 'Ham overcome by emotion when describing anguish', *Space Flight Now,* July 22, 2003. (http://spaceflightnow.com/shuttle/sts107/030722ham/ access date: 2/18/2004).

Harwood, W. (2003), *CBS News Satus Report* STS-107, 9:00 PM, 07/22/2003 (http://cbsnews.cbs.com /network/news/space/STS-107_Archive.html, access date: 2/18/2004).

James, D., P.D. Walker and D.J. Grosch (2004), 'SwRI ballistics tests help investigators determine the cause of *Columbia* loss', http://www.swri.edu/3pubs/ttoday/fall03/LeadingEdge.htm, access date: 2/18/2004; CAIB Report, vol. 1.

Kuhn, T.S. ([1962] 1996), *The Structure of Scientific Revolutions,* 3rd edn., Chicago: University of Chicago Press.

Langewiesche, W. (2003), 'Columbia's last flight: The inside story of the investigation, and the catastrophe it laid bare', *The Atlantic Monthly,* November 2003: 58–88.

Leusner, J. (2003), '*Columbia*: The final report', *Orlando Sentinel,* Wednesday, August 22, 2003.

Leusner, J., K. Spear and G. Shaw (2004), 'NASA avoids pinning blame for *Columbia*', *Orlando Sentinel,* February 15.

Mill, J.S. (1861), *Representative Government,* London: Parker, Son, and Bourn, West Strand.

Mill, J.S. ([1859] 1975), *On Liberty,* New York: Norton.

Orlando Sentinel (2003, February 13), 'O'Keefe makes probe more independent'.

Orlando Sentinel (2003, March 23), 'NASA managers missed chances to take closer look for Columbia damage'.

Orlando Sentinel (2003, July 1st), 'NASA managers missed chances to take closer look for *Columbia* damage'.

Pennington, H. (2003), 'We don't lack skill, just political will', *The Times Higher Education Supplement,* October 3, 2003.

Perrow, C. (1984), *Normal Accidents: Living with High Risk Technologies,* New York: Basic Books.

Pianin, E. (2003), 'Congress scrutinizing manned spaceflight', *Washington Post,* September 4, 2003.

Sawyer, K. (2003), '*Columbia's* "smoking gun" was obscured', *Washington Post,* August 24, 2003; CAIB Report, August 26, 2003, vol.1: 125, http://www.caib.us/news/report/ volume1/default.html, access date: 2/18/2004.

Schmitt, C. (1985), *Political Theology: Four Chapters on the Concept of Sovereignty,* trans. George Schwab, Cambridge, MA: MIT Press.

St. Petersburg Times (2003, September 4), 'Senators seek blame for NASA staffers'.

Thorpe, C.R. (2002), 'Disciplining experts: Scientific Authority and Liberal Democracy in the Oppenheimer case', *Social Studies of Science* 32: 525–62.

Turner, S.P. (2002), 'Scientists as agents', in P. Mirowski and M. Sent (eds.), *Science Bought and Sold,* Chicago: University of Chicago Press, pp. 362–84.

Turner, S.P. (2003), *Liberal Democracy 3.0*, London: Sage Publications.
Vaughan, D. (1996), *The Challenger Launch Decision: Risky Technology, Culture, and Deviance at NASA*, Chicago: University of Chicago Press.
Weber, M. (1975[1926]), *Max Weber: A Biography*, trans. and ed. by Harry Zohn, New York: Jon Wiley and Sons.

CHAPTER 7

FRANK NULLMEIER

KNOWLEDGE AND DECISION-MAKING

INTERACTION ANALYSIS OF POLICY ADVICE

Academic policy advice is usually analyzed at the level of structures: either the logics of science and politics are contrasted with each other or the procedures, organizations and institutions of policy advice are examined more closely. This paper, first, recommends a specific approach to the analysis of interactions occurring between scientists and politicians in policy consulting situations. Secondly, an extended form of structural analysis is suggested in order to overcome exclusive reliance on science and politics as analytical categories. Finally, the paper argues for the integration of interaction and structural analysis.

In order to analyze the interactive dimension of policy advice, elements of speech act theory – notably, John R. Searle's (1969, 1979) speech act typology and elements of Jürgen Habermas's (1999) recent attempt to base his concept of communicative rationality on speech act theory – may be used. Within German political science, speech act theory has so far been taken up by Katharina Holzinger (2001a, 2001b) in order to gauge the relative weight of argumentation and bargaining in mediation procedures. The following considerations aim to recommend the analysis of speech acts as a basic approach to the analysis of political interactions.

A speech act represents the smallest unit of verbal communication, in which a speaker performs an action in the presence of a listener (Austin 1962; Searle 1969, 1979; Krämer 2001). Each speech act consists of two elements, the propositional content and the illocutionary act. Whereas the propositional content indicates what the speech act refers to, the illocutionary act indicates which action is performed by uttering a statement. Where the analysis concentrates on the propositional contents of utterances, light is shed on the arguments, situational interpretations, causal relations and narratives advanced or suggested by actors. This path tends to be chosen by social scientific work under the heading of 'discourse analysis.' But the analysis can also concentrate on the action linked with a statement, thus focusing on the illocutionary component. The thoughts presented here exclusively concentrate on this type of analysis. Drawing on Searle's distinction of five types of speech acts – representatives (assertives), directives, commissives, expressives, and declarations, different forms of action performed with linguistic means can be classified and examined (Searle 1979). Moreover, Habermasian pragmatics provide a conception that high-

123

Sabine Maasen and Peter Weingart (eds.), Democratization of Expertise? Exploring Novel Forms of
Scientific Advice in Political Decision-Making – Sociology of the Sciences, vol. 24, 123–134.
© Springer Science+Business Media B.V. 2009

124 FRANK NULLMEIER

lights and substantiates the fact that each speech act is linked with validity claims (Habermas 1999).

Speech Act Theory and Interaction Analysis of Policy Advice

Academic experts do not only claim the truth of their empirical knowledge, but make a triple validity claim as to

- their empirical (descriptive, explanatory and prognostic) knowledge,
- their normative knowledge and value judgements yielded by it (evaluative knowledge),
- their vocabulary and conceptual tools (the dimension of 'world-disclosure' in philosophical terms).

Academic disciplines and subdisciplines, as well as particular theoretical approaches and lines of research within disciplines, are integrated by way of a stock of empirical-explanatory knowledge, a core of normative principles and a specific terminology. Academic experts participating in policy advice expect to secure the political acceptance of their knowledge in all three dimensions.

Such a perspective differs greatly from the familiar decisionist model that denied the existence of a genuinely scientific validity claim, at least in the area of normative knowledge. By contrast, the technocratic model considered the significance of empirical knowledge to be so far-reaching as to make normative questions irrelevant: the answers to questions related to the 'ought' and 'volition' were thought to be contained in empirical knowledge. In this model, then, scientists only make an explicit claim to the validity of empirical knowledge – due to its conclusive quality, the empirical validity of knowledge implies its normative validity.

Both models tend to ignore the role of *terminology*. Yet the conceptual aspects of world descriptions are in no way neutral with respect to empirical and normative validity claims, although the former are not merely a function of the latter either. The vocabulary of world-disclosure therefore represents – against the two models – an independent dimension of scientific knowledge that is brought into policy advice with a validity claim equal to the ones raised in the other two dimensions, all efforts to 'translate' notwithstanding.

But like any other communicating person, scientists offering policy advice link their utterances with yet another validity claim – namely, the claim of authenticity. No act of communication can get by without a reference to the presumptive authenticity of the speaker.

On the part of political actors, validity claims are effective, too, when communication in policy consulting situations takes place. Besides authenticity, there are likely to be claims related to the appropriateness of the used vocabulary and to the correctness of normative convictions. Specific empirical knowledge can be represented as true and valid as well. Hence there is no difference with regard to the number and kind of validity claims made by the two sides in the policy consulting situation. There is a basic symmetry of communication. How these implicit validity claims are dealt with is therefore of crucial importance:

KNOWLEDGE AND DECISION-MAKING

1. Validity claims can be consolidated, qualified, reflected upon, or shielded from questioning by others. A considerable part of communication attempts to either secure a particular claim or else, to question it. Many assertions are therefore accompanied by a meta perspective on the status of the validity claim advanced in them. A statement like "… at least that's what I have assumed so far…" can, for instance, contain a qualification or almost take back the claim that a stated piece of empirical knowledge is true. Such meta statements related to validity claims leave a characteristic mark on the interaction climate and are able to structure social relations. Whenever qualifying statements are only made on the political side, its inferiority in the area of empirical knowledge is maintained or suggested – scientists are thus potentially granted authority in matters concerning empirical knowledge, yet are, by the same token, under increased pressure to demonstrate and prove their superiority. Qualifying statements of inferiority can therefore be triggered by *distrust* as to the capacity of science to provide adequate messages with regard to the issues under consideration. Meta statements 'frame' validity claims in the sense of Goffman and hence open up the possibility of new games.

2. Validity claims are made in the context of speech acts. The particular type of speech act and validity claim that is made is crucial here. Distinguishing representative, directive, commissive, expressive and declarative speech acts according to Searle's classification, the illocutionary style of an act of communication can be inferred. If only the truth of empirical knowledge was debated in communication on policy advice, a dominance of assertive speech acts could be expected. Yet scientists frequently use statements with *directive character* in these contexts – more precisely, statements with a moderately or cautiously directive character: they ask for, suggest, or recommend the acceptance of specific bodies of empirical, normative and conceptual knowledge. Policy advice is thus by no means restricted to assertive or constative speech acts, i.e. statements on objects in the world. The directive character of expert statements can, however, create an asymmetrical situation between the participants. Whenever bodies of empirical knowledge are presented in the directive mode, their truth is no longer represented as being in doubt, and mutatis mutandis, the same holds for the correctness of normative or the appropriateness of conceptual knowledge. In the directive mode, the acknowledgement of presumptive knowledge by others as true, right or appropriate rather than their verification of this knowledge by way of arguments is aimed at. The promotion of specific pieces of knowledge and the creation of recognition can replace discursiveness in this mode.

In order to resist this outcome, politicians attempt to re-establish symmetry. They can be successful in achieving this goal by contrasting several divergent expert statements: the multiplying of directively presented knowledge ushers in a situation where directives are no longer only addressed to politics, but to representatives of science as well. The inflation of directive, yet contradictory speech acts decisively weakens their respective authoritative character. Asymmetry is offset by the inflationary use of this mode. The pluralization of scientific knowledge thus leads to a process of resymmetrization. Political actors can assume the role of observers of this acceptance game among scientists, of neutral referees and issue-oriented evaluators. However, this does not necessarily result in an argument-based assessment of alternative bodies of knowledge. Rather, by taking on the role of observers, political actors may indicate that they distance themselves from a knowledge 'show' and ultimately decide on the basis of aesthetic or rhetorical criteria which pieces of presumably valid knowledge will guide their actions. The conflictive relationship among participating scientists may also be used in order to confirm elements of one's own prior knowledge that are already presumed to be true or right. Thus the dangers of the directive mode are considerable. They must not be underestimated, all the more as directives can be considered as a type of speech act that is characteristic for political actors.

Speech-Act Theory and the Analysis of Political Decision-Making

Politics may, for our purposes, be defined as the totality of practices, procedures and institutions that are involved in the making of collectively binding decisions. As acts in which a collectively binding commitment to a particular course of action is made by or for a community, political decisions can be classified as *declarations* within the framework of speech act theory. Their form of existence depends on their self-declaration as decision. When a decision is communicated as such or declares to be a decision, a decision is made. Yet the destruction of alternative paths of action or the factual selection of a particular option is not yet a decision (although perhaps an act of choice). Only communication indicating the selection of a specific way to act, combined with communication related to the exclusion of other options, lets a decision become a decision. Hence a decision is the declaration of a decision (Wirth 2002).

The declarative form of decisions has to be distinguished from their textual content. (For the sake of the argument, let us imagine the written text of a resolution.) The text of the resolution can have a constative, descriptive content (assertives) if it includes an interpretation of reality, describes cause-effect relations, or clarifies the situational context of an action. Usually, however, these elements do not belong to the core of a decision, but instead reproduce the underlying assumptions about the context of a decision within the decision itself. Rarely, the content of a decision may also have expressive character, such as in expressions of gratitude, mourning, surprise or anger, etc. In acts of founding, on the other hand, decisions – themselves declarations – may also have a declarative content: Jacques Derrida has thus characterized the American Declaration of Independence as self-constitutive act of the American people in the form of an assertive statement (Derrida 2002). In general, however, the content of decisions represents commissives (binding the speaker as decision-maker) or directives (the speaker as decision-maker commits the hearer, the one affected by the decision, to something). Decisions often have the form 'We, the present, decide that X should be valid in the future' – an utterance whose status is as yet unclear and ambiguous from a speech-act theoretical perspective – or 'We decide that we (or otherwise, designated members of our group of decision-makers) want to do X in the future' – unequivocally commissive – or 'We resolve that others, actors A, B, C, ..., N (e.g., the executive) shall do something in the future' – a directive utterance. As decisions tend to have commissive or directive content, while consultants tend to act in the directive mode instead of obliging themselves to do something (except, e.g., to perform a new calculation until the next meeting), the *shared linguistic field of action* is directive. Scientists want to determine the behavior of others, and so do political actors. In politically and institutionally structured contexts, however, only elected political actors have the possibility to establish, by way of declarative acts, certain forms of behavior as truly binding in the future.

Political decisions are decisions that claim to be collectively binding within or for a politically constituted entity, such as the citizens of a political community and state. Bindingness has two sides. On the one hand, securing the obliging nature of a decision, once it is made, is at stake. This aspect certainly comes to mind first when the topic is discussed. Today, the bindingness of decisions is secured by way of legal codification. The bindingness of the law, in turn, depends on the degree to which legality alone proves sufficient to achieve this goal or else, legitimacy or mere social acceptance is required. Yet this post-hoc bindingness is closely linked with the process of *creating* bindingness – a process that is prior to the actual making of the decision. Engagement and the creation of obligations are needed in order to enable the commitment to a collective alternative of action communicated as decision. The commitment to a course of action that is *not* considered to be without alternative entails a series of engagements, binding the self and others, that eventually culminates in the decision itself. The political sphere is thus characterized by concern about the (1) creation and the (2) securing of collective bindingness. The fundamental interest lies (1) in the readiness to decide and (2) in the willingness (of particular groups within a community or of the entire community) to comply with these decisions. Processes of creating bindingness are often discussed under the heading of 'the capacity to impose decisions.' A proposed decision is thus linked with the claim to create and secure its collectively binding nature, in short: with a claim of bindingness. As long as the suggested decision is but a proposal, it aims to create bindingness; once the proposal has become an actual decision, the securing of bindingness comes to the fore. The factors that enabled or impeded the generation of collective bindingness can, however, always be reintroduced in communicative processes evolving around the securing of obligations linked with decisions. And in a similar vein, expectations regarding the (im)possibility of securing bindingness in the future can dominate the decisional phase itself. Politics aims to increase, in the interactions of policy consulting, the chances of creating and securing collective bindingness. At the same time, it tends to be aware of the one-dimensionality and problems of this yardstick. Whenever the creation of bindingness in the run-up to a decision is based on assumptions that foster the readiness to decide, but are found to be untrue after the decision, securing the ongoing bindingness of the decision becomes a political challenge. Therefore, politics is interested in truth (and correctness) if it has a sufficient time horizon and is interested in the durability of its power. Hence science is never – not even where interest in the securing of political bindingness dominates – just a resource of legitimation and an instance of post-hoc justification of a particular decision. In this perspective, a good outcome of policy advice is one that is able to simultaneously correspond to the benchmarks of normative correctness and empirical truth, and to the prerequisites of creating and securing collective bindingness.

A reference to politics as a process of creating collective bindingness could take up Jürgen Habermas's considerations and further modify the status of 'validity claims.' In the footsteps of Habermas, the term validity claim depicts *universal* or unconditional truth and correctness claims. Validity claims refer to *actor-independent* reasons for the validity of, e.g., a proposition. Because specific idealizations are immanent to every speech act, an empirical speech act can be criticized as non-universal and limited to *actor-centered* reasons. In the empirical study of political processes, on the other hand, claims of bindingness are the primary level of analysis, while universal validity claims are secondary.

The reasons that become politically relevant can be actor-independent or actor-related. In order to enable collective bindingness, it suffices for a group of actors who are to commit themselves to comply with a decision to collect enough actor-related reasons for it. The different scope of claims and of the kinds of relevant reasons – it is greater for political actors as they may refer to actor-related reasons as well – creates a fundamental asymmetry that may be experienced as difference between science and politics.

Perlocutionary Effects in Policy Advice

Yet whether the creation and securing of political bindingness can be analyzed exclusively at the level of verbal communication remains an important question – after all, the role of extra-linguistic – e.g., economic and military – power is well-known. Speech act theory offers the concepts of *perlocutionary effects* and *perlocutions* in order to capture extra-linguistic effects of a binding nature. Effects of speech acts are called perlocutionary if they could also be causally effected by way of non-verbal acts. Jürgen Habermas (1999: 126) distinguishes three classes of perlocutionary successes, in addition to perlocutions: in the first case, perlocutionary effects are the grammatical product of the content of a successful illocutionary act – an order is executed, a promise kept, an intention made real. In these cases, illocutionary objectives determine the perlocutionary ones. In a second class, perlocutionary effects are not regulated grammatically. These are the contingent result of a speech act that only obtains as a consequence of a successful illocutionary act – e.g., when a piece of news triggers joy or anxiety, or when an invitation to do something meets with resistance or not. The third kind of perlocutionary effect is at work if effects are reached in a way that is at first unrecognizable for the addressee, and the successful illocutionary act is but the basis for an unrevealed strategic action. Finally, perlocutions are characterized by the disappearance of this apparent precedence of the illocutionary objective as well. Overt insults and threats represent examples. Habermas's (1999: 134–137) revisions of his pragmatic theory of meaning (Habermas 1981, 1984) are therefore crucial for its adoption by political science: knowledge of the conditions for illocutionary success no longer guarantees the understanding of a speech act. In addition, knowledge of the conditions for perlocutionary success is required. Extra-linguistic power factors thus become constitutive for speech acts. The limitation to intra-linguistic features, namely linguistic (or linguistically represented) conventions, that the term illocutionary embodies, is replaced by a conception of speech acts that depends on the knowledge of potential perlocutionary effects.

The distinction of different classes of perlocutionary effects enables a more precise demarcation of strategic and communicative action. The boundary has to be drawn between the second and the third class. Extra-linguistic effects realized together with a successful illocutionary act leave the mode of understanding intact; only where a direct or hidden reference to an illocutionary act is missing, extra-linguistic effects become the mechanisms through which the transition to strategic action obtains. An interaction related to policy advice can therefore only be understood if the previous discussion, which concentrated on a classification based on the illocutionary sense of speech acts, is complemented by an analysis of perlocutionary effects. Where only first or second-order perlocutionary effects are present, communicative action is more or less dominant; if, on the other hand, third-order perlocutionary effects and overt perlocutions can be found, strategic action dominates. As soon as strategic action occurs, even occasionally, the danger arises that all actors change their mode of linguistic behavior and turn to strategic action in order to protect themselves. Yet in contexts of strategic behavior, linguistic understanding is subordinate to achieving one's respective goals. Interaction is reduced to the level of mutual observation, others are regarded as (potential) opponents, participants in the interaction try to influence each other instead of reaching a common understanding with regard to a particular issue, illocutionary intentions only matter as preconditions for perlocutionary successes, and illocutionary intentions are no longer pursued without reservations. In these strategic contexts, participants do not get involved in the mechanisms of communicative understanding without reservations, but rather, continuously reflect upon the effect that an utterance may have on political positions (of power). The mutual assumption of authenticity thus becomes obsolete, and all speech acts are deprived of their illocutionary binding force (Habermas 1999: 128). The truth values towards which actors are oriented are *not* transformed into truth claims aimed at intersubjective recognition. They rather become tokens in a strategic game.

Which type of speech acts dominates the interaction is only secondary in such a situation. What counts are the strategic intentions of the actors, who select and use the type of speech act and the illocutionary sense linked with it in an instrumental fashion. As a consequence, the speech act classification presented above has to be qualified and put into perspective. It has to be integrated in a structural analysis of the perlocutionary effects that accompany and contextualize speech acts. Simplifying to some extent, *understanding-centered and strategy-centered policy consulting situations* may be distinguished. Even where the content of scientific advice is directed towards the strategic goals of a collective actor – e.g., a party, communication between the persons participating in the interaction can take place in the mode of understanding – or else, in the strategic mode. A focus on understanding or strategy thus represents a quality of policy consulting situations and is part of the communicative 'style' of the politics-science relationship.

At this point, however, it is conceptually decisive to refrain from identifying science with an understanding-centered and politics with a strategic mode of communication. Let us further assume, for the purposes of this analysis, that there are two distinguishable actor groups among the persons involved in the interaction – i.e., scientists and politicians. The dominance of strategic communication may, then, be triggered by differences and conflicts among the politicians (1), among the scientists (2), between the group of scientists and the group of politicians (3), or between a faction made up of both scientists and politicians and another, similarly mixed faction (4). Whereas traditional analyses and theories of policy advice focus on the third kind of cleavage, possibilities (2) and (4) deserve at least as much interest today. Intra-scientific debates between lines of research and disciplines (case 2) can result in a departure from the mode of understanding in the internal communication of scientists. Perlocutionary effects may, for instance, slip in because research funding is at stake. One might speak of *science-induced* strategic behavior in this scenario. The formation of factions that cut across membership in the group of scientists or politicians is, according to firmly established results of policy research, an important empirical phenomenon (case 4). Such factions may already have existed before the beginning of the policy consulting situation or may have emerged in this situation; they can be rooted in conflicts among scientists and force politicians to take sides, but may also stem from a (party) political calculus that scientists accept due to their convictions, or which they find hard to ignore in the interaction. Science-induced strategic behavior therefore has to be examined as a subcategory of factional conflicts. Several conflicts may also arise simultaneously, creating an additional incentive for consistently strategic behavior. An interaction analysis of policy consulting bodies may attempt to draw on the dynamics of speech acts in order to reconstruct the four basic types of conflicts, their situative generation or reproduction, as well as their impact on the communicative style of policy advice.

Conversely, however, one has to keep in mind the possibility that attempts to maintain or introduce an understanding-oriented mode are initiated by political actors. They may fear the costs that arise when the validity of bodies of knowledge has been insufficiently clarified. Or they perceive a contradiction between the traditional understanding of science and the strategic nature of actually occurring communication. This may create the paradoxical situation that political actors demand a focus on understanding, supposedly the genuine mode of science, whereas the academic side pursues specific research-centered or political objectives in the mode of strategic action – in a perhaps cynical perspective on the gap between self-image and reality in the field of science. The 'naïve' view on the standards of science on the part of political actors, then, turns into a resource on the basis of which communicative action is called for. This example already demonstrates that acts of *meta communication*, as discussed above in the context of validity claims, play a crucial role for the development of communication and strategy-centered policy advice as well. The mode of communication can itself become an issue, thus either broadening the strategic discussion and proving the impossibility to circumvent it or else, creating the chance to push back perlocutionary elements at least temporarily. Whether understanding-centered policy advice takes place may be decided at the interactive level. Not only with regard to policy advice, but everywhere in the political sphere, situative 'islands of understanding' may emerge – depending on the behavior of participants. The latter, however, are embedded in structures that preconfigure every interaction in which they participate and therefore cannot be ignored by political science. Yet specific structures of science, politics or the science-politics interface per se by no means determine processes of policy consulting.

KNOWLEDGE AND DECISION-MAKING

STRUCTURAL ANALYSIS OF POLICY ADVICE

Beyond Duality

The relationship between knowledge and its impact – and notably, between knowledge and politics – tends to be debated as a relationship between politics and (academic) expertise today – currently, with a focus on commissions, policy advice, and expert bodies. The range of possible evaluations of these phenomena, especially from the perspective of democratic theory, is all too familiar: it stretches from the classical criticism of expertocracy and technocracy to the optimistic interpretation that an institutionalized expert discourse and a public discourse close to science may have an enlightening function, and from the positive evaluation of intra-scientific pluralization and the rise of counter-expertise to the current formula of deparliamentarization. A dual perspective on science and politics continues to dominate, and this paper has so far made no exception. Yet this dual picture was already criticized as outdated ten years ago by Renate Mayntz (1994), who instead stressed the significance of bargaining systems and the public:

1. In bargaining systems and networks, scientists turn into coactors and codecision-makers, and the situation of policy advice is only part of more complex bi- and multilateral communicative settings in which the involvement of science, among other issues, is determined. With reference to organized interests, pluralism is defined as the form of interest articulation in which a large number of organized interests compete and fight for influence on legislation, whereas the term corporatism is used where the state grants an almost monopolistic status to selected organized interests in order to delegate steering functions to institutionalized or informal arrangements that incorporate these interests. Along this line, one may, with regard to policy advice, discern a tendency of *expertocorporatism* and the trend to incorporate experts. Selected (male and, less often, female) scientists are given a role that corresponds to the role of important organized interests. They are involved in the business of government and entrusted with duties that would otherwise have to be performed by the government and parties. This inclusion of selected scientists changes their self-understanding. Scientists act as policy consulting entrepreneurs, as political *knowledge entrepreneurs*; they offer knowledge, stimulating and anticipating demand for it against the backdrop of mutual adaptation processes on the supply and demand side.

2. Whereas in the past policy advice was often given in the arcanum of politics, it is out in the public today. Information about who consults whom is known in the expert public and made known by political reporting in the mass media. Moreover, policy advice is directly addressed to the public rather than to specific political actors. Policy advice and information of the public are, to some extent, merged. Renate Mayntz, rather optimistically, refers to the latter trend as the transition from policy to societal advice. Yet the increased significance of the mass-media public for the political impact of knowledge must be looked upon in a less enthusiastic fashion today. Political impact is more and more created

through the public presence of experts who engage in the promotion of their knowledge and therefore become coresponsible for the justification of policies. It thus becomes imperative for knowledge to create attention – in the sphere of politics, but for the purpose of intra-scientific gains as well. The competition among disciplines and bodies of knowledge can increasingly be fought out through media presence. Knowledge entrepreneurs in the field of policy advice create an impact on the disciplines they represent through their sheer media presence. They can thwart or strengthen media strategies that appear to make sense from a disciplinary perspective. Academic media stars thus have the potential to leave a mark on their disciplines and to become actors in the political game. The media certainly have become a core arena of policy advice. Yet media presence does not equal societal advice, and the arcana of policy advice continue to survive.

Policy Advice as Profession

As elsewhere, the developments of the last couple of years have led to the emergence of a specialized sector of policy advice in the Federal Republic of Germany – a sector located somewhere between politics, consulting and science. Consulting firms establish new departments of public affairs, university graduates with social-science degrees start policy consulting firms. Foundations build up think tanks, political managers formerly employed by the parties enter the business on their own. Policy advice may understand itself as a broader form of lobbying for organized interests, especially corporations, but also as a new form of activity that delivers data and concepts from an independent vantage point that is nevertheless close to the political and academic spheres. This new field is, however, characterized by the fact that it is not strongly anchored in the scientific system. The new type of policy consultant is a competitor of traditional academic policy advice (Tiemann 2004), as diminished academic status is compensated by greater proximity to the political sphere and a more pronounced practical orientation. This new sector that devotes itself specifically to policy advice has to be examined in order to discern whether it develops according to a scientific, political or economic logic. There is considerable evidence for the thesis that a market-oriented logic must prevail, as permanent state funding is lacking. The professionalization of policy advice and the growth of a specialized sector comprising knowledge firms, think tanks and policy consultants are likely to result in the commercialization of a field once dominated by science. The development of this hybrid field will therefore probably follow a different logic than the ones prevailing in the neighboring fields of politics and science.

The dual perspective, which has appeared too narrow ever since the emergence of network structures, thus has to be replaced by an analytical triangle or even a square of actors: politics, science, the policy consulting sector (and fourthly, the media). As a consequence, the interfaces between science and policy advice (1) and between policy advice and politics (2) emerge as new problem areas:

1. Recent political debates illustrate the fact that consultants (in the sense of corporate consultants) can no longer be distinguished from the suppliers of scientific

advice. Policy advice is perceived as one large and interdependent or even as a unified field in the media public. The membership of scientists in commissions and the contracts of consulting – now including policy consulting – firms appear to fall in the same category. As a consequence, securing the reputation of scientists engaged in policy consulting within the scientific system itself becomes problematic. Participation in exercises of policy advice can result in the decoupling from science or else, trigger a new wave of differentiation: the formation of policy consulting branches in each discipline. A harmonious relationship of neighboring fields may result from this trend, but it may also result in a confrontational relationship in which the policy consulting branches see their status questioned. Academic policy advice and policy advice in the style of consulting firms do not yet fully converge, but there is a rapprochement, especially in the field of administrative and organizational reform. Similarly, hybrid developments linking science with (policy) consulting, such as a new branch of political management, can be imagined.

2. With the policy consulting sector, a commercially oriented actor has entered the stage, and institutions of academic policy advice receive more public attention than in the past. Politics therefore has to consider which are the adequate forms of policy advice today. Policy advice has become a normal element of the political system in the area between the system of interest representation and articulation (parties, organized interests, associations, social movements, the politically informed public) and core institutions (especially, parliaments and the executive branch). There may be attempts to institutionalize this sector (e.g., national academies of science), to regulate it (e.g., by way of 'policy advice legislation' along the lines of party and never-passed organized-interest legislation, stricter rules for consulting contracts, or – at the very least – a 'code of ethical conduct' for policy consultants), and to build on the survival of different types and approaches of policy advice. In any case, decisions on policy advice as necessary components of each political process have to be constantly renewed. Scientists' complaints about the loss of influence of academic policy advice thus tend to underestimate the complexity of the new situation. Attempts by some scientific actors to regain authority through better organization represent, politically speaking, nothing but a strategy of organizational consolidation and corporatism for the field of science that aims to predetermine political decisions by way of policy advice.

INTERACTION AND STRUCTURE

The macrostructural process of differentiation that has led to the emergence of a specialized policy consulting sector is certain to have consequences for the *interactive structures* of policy advice. A first hypothesis related to the more economically structured form of policy advice that is likely to emerge consists in the expectation of a rising frequency of commissive speech acts, as they represent acts of self-commitment in the sense of producer behavior. However, this hypothesis may be as false as the hypothesis suggesting that assertive speech acts dominate in science. It may be more plausible to expect that the communicative mode of policy advice will increasingly be characterized by strategic behavior. The process of commercialization also implies a growing significance of third-order perlocutionary effects. A third hypothesis is linked with the observed emergence of a policy consulting sector and the end of the science-politics dualism: reflections and meta communication will become more frequent in the public and in policy consulting situations themselves. The eroding authority of science and the rise of economically interested consultants will make more frequent negotiations on the status of utterances imperative. So much for some hypotheses on changes at the interactive level that may follow structural change. Only an empirical analysis can answer the questions raised here. A productive way of dealing with the increasingly complex nature of policy advice in the world today will only succeed if a renewed and extended structural analysis and more sophisticated methods of interaction analysis in empirical research are combined and made to complement each other.

Universität Bremen, Germany

REFERENCES

Austin, J.L. (1962), *How to Do Things with Words*, Cambridge, MA: Harvard University Press.
Derrida, J. (2002): 'Unabhängigkeitserklärungen', in U. Wirth (ed.), *Performanz. Zwischen Sprachphilosophie und Kulturwissenschaften*, Frankfurt a.M.: Suhrkamp, pp. 121–8.
Habermas, J. (1981), *Theorie des kommunikativen Handelns. Volume 1: Handlungsrationalität und gesellschaftliche Rationalisierung*, Frankfurt a.M.: Suhrkamp.
Habermas, J. (1984), *Vorstudien und Ergänzungen zur Theorie kommunikativen Handelns*, Frankfurt a.M.: Suhrkamp.
Habermas, J. (1999), *Wahrheit und Rechtfertigung. Philosophische Aufsätze*, Frankfurt a.M.: Suhrkamp.
Holzinger, K. (2001a), 'Kommunikationsmodi und Handlungstypen in den Internationalen Beziehungen. Anmerkungen zu einigen irreführenden Dichotomien', *Zeitschrift für Internationale Beziehungen* **8** (2): 243–86.
Holzinger, K. (2001b), 'Verhandeln statt Argumentieren oder Verhandeln durch Argumentieren? Eine empirische Analyse auf der Basis der Sprechakttheorie', *Politische Vierteljahresschrift* **42** (3): 414–46.
Krämer, S. (2001), *Sprache, Sprechakt, Kommunikation. Sprachtheoretische Positionen des 20. Jahrhunderts*, Frankfurt a.M.: Suhrkamp.
Mayntz, R. (1994), 'Politikberatung und politische Entscheidungsstrukturen: Zu den Voraussetzungen des Politikberatungsmodells', in Axel Murswieck (ed.), *Regieren und Politikberatung*, Opladen: Leske + Budrich, pp. 17–29.
Searle, J.R. (1969), *Speech Acts: An Essay in the Philosophy of Language*, Cambridge, UK: Cambridge University Press.
Searle, J.R. (1979), *Expression and Meaning. Studies in the Theory of Speech Acts*, Cambridge, UK: Cambridge University Press.
Tiemann, H. (2004), 'Im Dickicht der Beratung: Politik und Wissenschaft im 21. Jahrhundert', *Zeitschrift für Sozialreform* **50** (1/2): 46–50.
Wirth, U. (ed.), (2002), *Performanz. Zwischen Sprachphilosophie und Kulturwissenschaften*, Frankfurt a.M.: Suhrkamp.

CHAPTER 8

WILLEM HALFFMAN AND ROB HOPPE

SCIENCE/POLICY BOUNDARIES: A CHANGING DIVISION OF LABOUR IN DUTCH EXPERT POLICY ADVICE

The tasks science-based experts perform for policy are many. In the traditional set of instrumental tasks, experts provide factual information to policy makers, assess future policy outcomes, or determine effects of past policies. However, the practice of policy expertise is much more varied. Experts may criticise policy makers' problem definitions, redefine problems, reframe policy beliefs, point at unanticipated outcomes, suggest alternative strategies, interpret policy and provide critical reflection, or even mediate in controversies (Renn 1995; MacRae and Whittington 1997; Bal et al. 2002).

This does not imply that experts do, or should do, all of the above all the time. There is no universal list of experts' tasks. Policy makers may prefer to rely on their own knowledge, their own mediating skills, or their own ability at critical reflection. Especially in times of difficult political bargaining, 'critical reflection' is the last thing politicians want, especially from the experts. In other cases, the very status of the expert is at stake and actors may attempt to redefine what can be considered a matter of expertise and what a matter of policy. That is why we can analyse the relation between experts and policy makers as a complex and contested division of labour. This division of labour consists of a *boundary* that demarcates who can and cannot be considered an expert in various degrees, and articulates the coordination between actors who have come to be considered 'experts' and 'policy makers.' Such boundaries are the outcome of – and form the resources for – continuing *boundary work*, the further articulation, reproduction, or modification of this division of labour (Jasanoff 1990; Shapin 1992; Gieryn 1995; Gieryn 1999; Halffman 2003).

Over time, patterns have developed in this division of labour, varying between countries and policy sectors. Some advice giving tasks come to be recognised as important, and some as the job to be fulfilled by experts exclusively. Accordingly, the process of providing expertise is organised in different formats, ranging from ad-hoc expert committees, consensus conferences, contract research, to even informal meetings in a personal network. Hence, practices of advice giving develop into institutions, i.e., more or less routinised patterns in which expertise and policy are demarcated and coordinated.

135

Sabine Maasen and Peter Weingart (eds.), Democratization of Expertise? Exploring Novel Forms of Scientific Advice in Political Decision-Making – Sociology of the Sciences, vol. 24, 135–151.
© Springer Science+Business Media B.V. 2009

Governments have installed expert organisations for the specific purpose of advising policy. Such organisations develop a body of knowledge, formal and informal rules about how to provide advice, a more or less guaranteed budget, or a conception of what is and is not their business. Government departments have developed procedures for commissioning research, ranging from model contracts to informal routinised practices of commissioning expertise. Scientists also have developed institutions for expert advice giving, such as professional codes of conduct, networking platforms for meeting policy makers, or conceptions of what kind of public roles their profession should or should not play (Peters and Barker 1993; Hoppe 2002a).

Most of the descriptions of this institutionalisation of science/policy boundaries tend to homogenise their account in one of two ways. The first and static homogenised account describes patterns in science/policy boundaries as matters of national style. Such accounts attempt to identify a typical pattern in a country, and possibly relate this to crucial historic events (such as the imposition of the Code Napoléon), or the development in key macro institutions (such as the legal system or the civil service). For example, the US is typified as having an adversarial style of expertise, which is then related to an adversarial legal system and majoritarian politics. Such typifications are strong at accounting for the mutual connectedness of the institutionalisation of expertise and policy, the co-production of science and policy, but tend to have difficulty in accounting for short-term changes in the organisation of expertise or for the diversity between policy sectors (Brickman et al. 1985; Vogel 1986; Bakker and Van Waarden 1999; Renn 1995; Halffman 2003).

The second and dynamic homogenising conceptualisation is that of the grand transition. Such accounts try to identify how science/policy boundaries are changing from one form to another, compensating for the static bias of the national style notion. For example, transition accounts will point at increased transparency and accountability of experts towards citizens, increasing possibilities of wider participation in the production and evaluation of expert knowledge claims. Similarly, binary notions such as mode 1/mode 2 science (Gibbons 1994) or normal/post-normal science (Ravetz 1999) point at such transitions. Once again, the account tends to homogeneity: one state of affairs leads to another in an encompassing grand narrative.

In this chapter, we argue that these homogenising accounts of science/policy boundaries fail to address the diversity of institutional patterns, as well as the wide-ranging ideological disagreements that form their backdrop. With very inductive empirical accounts of the development of public expertise in the Netherlands over the last decades, we will show how at least three patterns of science/policy boundaries can be identified: a corporatist, a neo-liberal, and a deliberative pattern. In doing so, we want to acknowledge the importance of the connectedness of expert and political institutions of the national styles-approach, but, while acknowledging the importance of national macro-institutions, bring forward and make sense of the maelstrom of transitions in the organisation of public expertise. We will show that various patterns continue to exist next to each other in Dutch national expert institutions; that the tension between these patterns is loaded with ideological disagreement and contradiction; and that we find diverse processes of change rather than one transition.

SCIENCE/POLICY BOUNDARIES

CORPORATIST EXPERTISE: THE PLANNING BUREAUS AND THE ADVISORY COUNCILS

In corporatist policy arrangements, a restricted set of what are considered a sector's main policy actors are formally accredited to participate in the policy arena. In various forms and with considerable variation over time, (neo-)corporatist modes of decision making have been strong in the Netherlands, especially in socio-economic policy. In these modes, the institutionalisation of expertise takes one of two typical forms. In the first form, the formally accredited actors mobilise their own expertise. In the more technical negotiations, actors may even be represented by experts. For example, a university professor may participate in a negotiation over health insurance benefits to represent the position of patients. We see this pattern strongest in the old system of Dutch national advisory councils. In the second form, the experts draw up the playing field for the corporatist negotiations. Experts then act as the linesmen of politics, indicating within which constraints actors can operate. Just like in the soccer game, they wave a flag whenever the negotiation game exceeds budgetary constraints or becomes unrealistic about next year's economic growth. This pattern can be found most clearly in the present position of Dutch planning bureaus.

To start with the former, the Dutch corporatist tradition of ruling by consensus among an elite of 'relevant actors' (the model of recognised employer organisations and unions expanded to other sectors of society) had led to a large number of sector-specific advisory councils. By 1976, there were 402 of them, providing platforms for negotiation and attempts to build sectoral consensus. Not all of these were strictly advisory, as some even had tasks in policy implementation or regulation. The unbridled expansion of advisory councils eventually became a policy issue itself, resulting in a long list of reports. Over a period of two decades, various reports advised how to reduce their number and create some order. Meanwhile, the advisory bodies themselves changed. Whereas an inventory of 1976 had counted a third of the membership of these councils as 'independent experts' (Wetenschappelijke Raad voor het Regeringsbeleid 1977), by 1993 this had become two thirds (Oldersma 2002). The participation of various groups in 'their' policy sector was increasingly taken over by experts representing their position.

This tendency to professionalise, the shift from interest representation to interest-cum-knowledge representation, was eventually taken to a radical conclusion. At the initiative of the Ministry of the Interior, the debate resulted in two laws in 1997, one providing a new framework for advisory councils and one abolishing nearly all of the existing ones. After this radical reorganisation, there were only eleven major advisory councils left, next to about seven highly specialised ones. With the exception of one, the Social and Economic Council, all the advisory councils were now considered expert councils. They were to advise with knowledge rather than interests. Another key principle was that advisory councils were to break out of their policy-specific niches, ranging across policy sectors. Advisory boards were to become more general and less tied to the specific interests and perspectives of the traditional policy fields. However, given that these advisory councils resort under the responsibility of individual government departments and that some of them have very specific functions in policy, most advisory councils have remained sector-specific, although sectors have come to be defined somewhat wider than before. The logic of diverse government departments, each with their specific professional strongholds, style of operation, and sectoral networks, proved stronger than that of a legal reorganisation (Klink 2000).

To introduce a second major development, the Dutch planning bureaus provide government departments with assessments of the state of affairs and of future developments in their policy sector and relate these to policy options. The term 'planning' is somewhat misleading. They hardly ever 'plan' in the sense of selecting goals and allocating means, but rather analyse and forecast. The planning bureaus' status in the Dutch polity is exceptionally commanding, to the degree that the environmental and especially the economic planning bureaus routinely assess likely outcomes of political parties' programmes prior to elections.[1] Political parties who refuse to submit their programme to such an analysis find their position severely undermined. Even the presentation of an uncertified oppositional counter-budget in Parliament, as an alternative to the government's annual budget, can be a political liability (Van den Berg et al. 1993; Centraal Planbureau 2003a). For purposes of political negotiation and bargaining, predictions of economic growth, budget shortages, or unemployment figures are in most cases accepted as true and unproblematic inputs for decision making. In addition, government is even legally required to consult planning bureaus at some points in the policy making process and in the annual budget cycle. As such, the planning bureaus occupy positions as obligatory passage points for Dutch politics that would be considered unacceptably technocratic in most other countries (Van den Bogaard 1998).

By 2002, there were planning bureaus for economic policy (Central Planning Bureau, CPB, established 1947), social and cultural policy (Social and Cultural Planning Bureau 1973), environment (the Netherlands Environmental Assessment Agency in the National Institute of Public Health and the Environment, RIVM, which received the 'planning bureau function' officially in 1996), and urban and regional planning (the Netherlands Institute for Spatial Research,[2] lifted out of the Department of Spatial Planning in 2002). The casual use of the term 'planning bureau function' now suggests that planning bureaus are an entirely natural phenomenon, a logical part of policy making: the 'function' needs to be fulfilled. However, the planning bureaus followed only one particular model for organising expertise, that of the Central Planning Bureau. Its strong reputation for econometric modelling, high policy impact, and close ties with the Ministries of Economic Affairs and Finances brought other ministries to develop competing expert resources of their own for the departmental tug-of-war and formed an enviable status for other expert organisations.

With four official planning bureaus in place, there are many opportunities for tension. Advisory organisations, like professional organisations, survive by claiming specific areas of expertise or specific approaches that make them unique and worthy of collective funds in the ecology of knowledge (Abbott 1988). Such strategising can take the form of competition, as different government departments pitch organisations and their reports against each other in the heat of political conflict. However, the planning bureaus seek their legitimation in 'independence,' in a 'neutral,' or at least 'third party' stance with respect to the political process. In the presentation of hard figures, open competition is a high-risk strategy. It could easily lead to deconstruction of facts and mutual undermining of authority. In recent years, the planning bureaus have tended towards a strategy of accommodation, seeking mutual coordination through consultation. One example is the Planning Office Directors Consultative Committee, an informal structure negotiating the relations between the planning bureaus, but also articulating what it means to be and act like a planning bureau (Overleg Directeuren Planbureaus 1996; Wetenschappelijke Raad voor het Regeringsbeleid 2001; Centraal Planbureau 2003b).

The term 'independence' of the planning bureaus thus has a specific meaning: planning bureaus claim that blatant political influence will not alter their advice, even if unwelcome. However, 'independence' clearly has its limits, which is acknowledged by the planning bureaus themselves. For example, research agendas are coordinated with long-term policy perspectives and members of planning bureaus are often present as advisers at top-level policy meetings (Centraal Planbureau 2003c). In some cases, directors of the planning bureaus will even attend Cabinet meetings (Hoppe 2002b). This is not surprising, as these are exactly the kinds of 'dependencies' that assure a productive cooperation between experts and policy makers. To enact 'independence' it needs to be articulated, specified in practices, rules, and institutional arrangements. Allowable dependencies need to be distinguished from unallowable ones; the organisation must be kept out of the vortex of mediated politics, and where it enters this vortex its image must be spun with care.

In the attempt to create independence, there is some preoccupation with the organisational status of planning bureaus. The 'closeness' to their respective government departments tends to be understood as a matter of organisational schematics. Over the last couple of years, the newly preferred organisational status for the planning bureaus has been that of an agency, formalising the arm's length position of planning bureaus. One of the key issues is the diversification of clients, since agencies normally do not work for government alone, but are expected to raise some of their own money on the contract research market. Presently, in the case of the RIVM, these clients are explicitly not to include industry (unless if requested by a government department) and the common planning bureau protocol also states that "commercial research assignments are generally seen as a threat to the credibility / independence of the planning offices" (Overleg Directeuren Planbureaus 1996; Rijksinstituut voor Volksgezondheid en Milieu 2003).

These shifts thus represent ambivalent changes in the division of labour in corporatist patterns. Planning bureau assessments are not always received without objection and not all planning bureaus have achieved the status of the economic one, the CPB. However, the degree of acceptance of assessments as reliable, independent, and for all practical purposes 'true,' is remarkable. In spite of the fact that planning bureaus are seen as resources by various government departments, positioning them as ways to promote their policy agenda in their mutual competition, planning bureaus' identification of expected policy outcomes tends to be widely accepted, thus creating the space within which bargaining is possible. Similarly, the advisory councils have moved from a logic of interest-cum-knowledge representation to one of representation of the issues and the state of 'relevant' knowledge. This does not mean that the restriction of policy access to the major actors, typical for corporatist decision-making, has disappeared. Rather, experts have been repositioned, providing especially the executive with stronger leverage to break through corporatist deadlocks (Hemerijck 1994).

Goals and preferences may vary, but in a political system that is highly diverse, with complicated coalition governments, having stern arbiters that judge the outcomes of proposed policies reduces the complexity of negotiations. In a fluctuating multi-party system where results of elections are always politically ambiguous and with strong tendencies to pacify conflict (rather than to humiliatingly defeat an opponent who could be a needed ally tomorrow), experts are welcomed as linesmen of politics. Especially in the case of planning bureaus, their verdict is accepted even if this means that some policy alternatives are blocked off by assumptions in a computer model that are technical and hard to question.

NEO-LIBERAL PATTERNS IN PUBLIC EXPERTISE

The development towards a (neo-)corporatist linesman of politics in Dutch expertise is only one. Other developments point in the direction of a growing importance of a neo-liberal pattern for the organisation of the science/policy boundary. Typical of this pattern are the small state philosophy, leading to the 'externalisation' of expertise out of government departments, and a strong emphasis on the market to coordinate expert resources. We already indicated that most planning bureaus have moved towards agency status. In addition, the erosion of the old corporatist advisory system at least ran parallel to the neo-liberal rejection of corporatist policy arrangements, where state and society are seen as unacceptably colluded. There are two more indications that a neo-liberal pattern of public expertise is becoming more important in the Netherlands: the radical externalisation of expertise at some ministries, and the growing contractualisation and commodification of expert knowledge.

Most government departments are entangled in a continuing struggle to find the most suitable position for expertise. While there is an increasing awareness of the importance of knowledge for successful policy making, and while concepts such as 'knowledge intensive administration,' 'knowledge infrastructure,' or 'evidence based policy' are increasingly popular among analysts (Beker et al. 2003; Paardekooper 2003; Dijstelbloem and Schuyt 2003; Kronje 2003), it is by no means clear what this means for the organisation of the science/policy boundary. This struggle is a matter of both the internal departmental organisation and of external relations with expert organisations. It is in this context that neo-liberal solutions have surfaced most radically.

First, is it preferable to have one research division in a department, or is research and knowledge to be managed within the various functional units of a department? The organisation of expertise in a department can have considerable consequences for the freedom sub-units may have to gather and supervise expertise. Distributed control over research may allow expertise to be fed into policy more directly, as long as the civil servants concerned manage the process well. A centralised research directorate, on the other hand, may be more apt at guaranteeing quality of research, at the potential cost of developing into an ivory tower. Various departments opt for different solutions here. For example, since about 1995, the ministry of Social Affairs and Employment toyed with the idea of a specialised research division. Proponents argued that it would lead to better and more shared knowledge. After much deliberation, a research division was indeed set up, only to be abolished again a year later. The ministry now focuses its knowledge management on permanent learning for its staff through an 'Academy,' in correspondence with the 'lifelong learning' policy for the Dutch work force (Ministerie van Sociale Zaken en Werkgelegenheid 2003). In contrast, the department of Traffic and Water Management has experimented with a research division since the early nineties, which has resulted in a special Directorate for Knowledge and Development (Ministerie van Verkeer en Waterstaat 2003).

However, secondly, departments have struggled with the question to what extent expertise can be externalised into more or less autonomous agencies – or should be outsourced to consultancies or other research organisations, among which universities. Over the last decade, some departments have externalised expertise in a quite radical way. The form of externalisation has varied and not all departments have followed suit. The department of Spatial Planning has kept part of its division of planning, but has created the Spatial Planning Bureau. The department of the environment continues to grant more autonomy to the RIVM, first changing annual agreements on research to bi-annual ones, and eventually giving the institute agency status in January 2004.

Externalisation comes with specific problems, which lead to new coordinating institutions. The Ministry of Health, Welfare and Sports has virtually lost all its short-term influence over research, as the major research institutes under its responsibility work only with longer term research programmes (such as the Social and Cultural Planning Bureau and the health research resources of the Dutch national fund for scientific research) (Beker et al. 2003). In addition, this ministry has invested heavily in the development of independent 'knowledge centres.' This may lead to increased availability of knowledge in the policy sector, but most of these centres tend not to see government as their main client (see below). The Ministry of Agriculture, Nature and Food Quality has organised its expertise in a separate Expert Centre. The sole client of this Centre is the Ministry and it is to prepare and evaluate policy, as well as critically follow it. Nevertheless, the Ministry has its own department of Science and Knowledge Transfer, which is to operate as a knowledge broker (Expertisecentrum LNV, 2003a, De Wit 2003; Ministerie van Landbouw 2003; Expertisecentrum LNV 2003b). The creation of such a knowledge broker seems typical. An increased distance may generate a stronger resource for building legitimacy, claiming 'independence' of expertise, it also creates a gap between immediate policy needs and the agenda of professional researchers. This induces complex negotiations over mutual relations and degrees of control over research agendas.

There may not be a single dominant pattern in the organisation of expertise within departments or in the degree of externalisation of research, but to the extent that research is externalised, there does seem to be a new development of complex and increasingly formalised negotiations over research projects and programmes. From the perspective of researchers, this development can be seen more clearly. Dutch government departments tend to keep an active role in the oversight of commissioned research. For example, the use of advisory committees has become standard practice. These committees are typically comprised of civil servants with some expertise in the matter at hand, as well as experts, usually sympathetic to the project. Advisory committees are consulted over the problem formulation at the beginning of a project, over the progress of projects, and over the end results. Careful selection of an advisory committee is acknowledged as a key instrument for maintaining control over a research project and for guaranteeing that the results will be of use for policy. In combination with financiers claiming ownership over reports and with disclosure clauses, government departments can exert considerable influence over the formulation of published reports and over the timing of their publication. In two recent cases that made the national newspapers, government departments used these instruments to prevent discussion of research results among scientists, rewrite conclusions or recommendations, or delay publication of unwelcome news (Ramdharie and Trommelen 2003; Schreuder 2003).

Who gets commissioned? Although there are only very rough indicators of how government departments spend their research resources, it does seem clear that universities are slowly slipping out of policy makers' favour, at least with national civil servants looking for policy advice. This follows the image portrayed by civil servants that departments presently favour either the authority of an established planning bureau or the convenience of a consultant – although universities are still good enough for €156 million's worth of commissions every year. In many cases, policy makers see academic researchers as unpredictable, over-principled and as refusing to stick to the policy problem at hand. Especially in an instrumental approach to researchers as fact-finders, consultants may provide more suitable avenues. From their perspective, some Dutch social scientists have complained that the grip on commissioned research has become too tight, sometimes verging on manipulation (Köbben and Tromp 1999; Sociaal Wetenschappelijke Raad 2000).

The amount of research government commissions to universities is relatively small in relation to their total research funds. The €156 million come out of €2.278 million in 2000 (Centraal Bureau voor de Statistiek 2003). Nevertheless, with 7%, this is still considerable and especially so for the social sciences and the unknown share they manage to obtain from it. In the second half of the nineties, university budgets have stagnated. Many social sciences, such as sociology, have been faced with reduced numbers of students, partly for demographic reasons[3] and partly because of the popularity of new programmes such as communication studies. Contract research may offer some extra oxygen for research groups. However, either at the initiative of individual researchers or of universities themselves, much of the most profitable research has been organised outside of the restrictive corset of university organisations, of academic peer pressure, and of civil servants' employment regulations which still protect academics. Nevertheless, in the case of the social sciences, government is the main and sometimes the only client. Even though universities may no longer be as important to civil servants for advice, civil servants have become all the more important to academic social scientists looking for extra funding.

Here too, we see commodification combined with a need for new forms of coordination. If there is a general line in the development of how government departments gather expertise, it is that of contractualisation: relations have become increasingly formalised and legalised. The control practices around commissioned research have refined, as in the practice of setting up advisory committees. In addition, some departments have developed guidelines on how civil servants should set up commissioned research, often accompanied by sample contracts and detailed rules. Even the relation between departments and their research institutes now has taken a contractual turn in which agreements are made on research programmes and targets. The internationalisation of the market of expertise and the increased competition between sources of expertise are likely to drive this development further. The contractualisation of research supports civil servants in setting up advisory relations with a wider range of researchers, even researchers or consultants who are not familiar. In spite of such advantages, the hidden transaction costs are considerable, for example in the form of legal overhead, over-instrumentalised knowledge, problematic accumulation of knowledge over time, or of new institutions to deal with newly produced problems of science/policy coordination.

Shifts Towards Deliberative Patterns of Organising Public Expertise

In deliberative conceptions of democracy, public reasoning and discourse are seen as crucial aspects of politics. Therefore, we call a deliberative pattern of public expertise all those forms of organising the science/policy boundary that position expertise as a collective resource in public debate, wherever this takes place (parliament, sectoral forums, media). The pattern is frequently connected with discourses of public participation, the importance of experiential knowledge, public accessibility of knowledge, and reflexive awareness of the possibilities and limitations of expertise (Hajer and Wagenaar 2003), thus contrasted with the restrictive nature of corporatist structures and the primacy for strictly representative democracy. We find the pattern most clearly in the experiments with interactive expert decision making stimulated by the Rathenau institute, the new phenomenon of 'knowledge centres,' the improved self-understanding of public expert organisations, and – much more ambivalently – in the expanded expert resources of the Dutch parliament.

Traditional corporatist patterns of expertise to a certain extent did take into account opposing views in matters of expert knowledge. However, these patterns operated around relatively rigid corporatist channels of representation. The typical new issues of the risk society create new collectives, for which corporatist models prove insufficiently flexible (Beck 1992 [1986]). For example, the initial attempts at regulatory negotiation and corporatist mediation in environmental issues, with the department of the environment stimulating the development of environmental groups through subsidies and the construction of statutory advisory boards in the eighties, gave way to new models in the nineties. This included government addressing societal actors directly, environmental groups negotiating with individual companies, or the construction of non-governmental regulatory bodies, such as for eco-labelling of food. In this setting, it is never a priori clear where relevant expertise will come from. New actors appear around new policy issues, bringing their own knowledge or their own concerns about how policy expertise is framed.

A key actor behind new ways of handling expert knowledge in controversial issues that defied traditional policy making, was the Rathenau Institute. In its attempts to find ways out of the conundrums of 'impact' type of technology assessment and to clarify its position towards government, the Institute started to experiment with new forms of bringing experts, citizens, stakeholders and policy makers together. It copied elements of the Danish consensus conferences or of American citizen juries, but also thoroughly stimulated the innovation of interactive and constructive technology assessment. The methodology developed by the institute is slowly finding its way to other organisations, for example in transition management aiming for sustainable technology (Van Est et al. 2002).

Another interesting example of the deliberative pattern of organising expertise is the emergence of 'knowledge centres,' since about 1998. Although quite a few research institutes have simply relabelled themselves to catch the buzz word, the new knowledge centres claim to be qualitatively different. They claim to make knowledge more available for policy use, either by integrating knowledge, simply accumulating knowledge, or by performing a role as knowledge broker. Knowledge centres are seen as facilitators of a collective and public learning process, targeted at practitioners in general, rather than governmental policy makers in particular (Beemer and Den Boer 2003). Their organisational form ranges from merely a portal web site, run by a handful of people, to the research facilities of an entire university (Wageningen). There are currently about 115 knowledge centres, largely funded publicly (Ketting 2002). Knowledge centres generally organise themselves around policy fields or specific policy issues, rather than around the traditional definitions of research fields or disciplines. For example, there are knowledge centres for sustainable building (Nationaal Dubo Centrum 2003) or urban policy (KEI Kenniscentrum Stedelijke Vernieuwing 2003; Kenniscentrum Grote Steden 2003). Especially the Ministry of Health, Welfare and Sport has actively supported the creation of knowledge centres, leading to a boom in this area.

What is new about the knowledge centres is not so much the claim of improved knowledge transfer, but their post-professional positioning, outside of those major strongholds of disciplines, universities, as well as established research institutes. This means knowledge centres can – in principle – operate across research fields and professional jurisdictions, integrating knowledge on an issue-basis in various forms of inter-disciplinarity. For example, the Knowledge Centre for Large Cities is setting up initiatives cutting across the division of labour between the social and physical aspects of urban planning (Kenniscentrum Grote Steden 2003).

Evidently, this raises questions about quality assurance. Professionalised knowledge may have the bad reputation of becoming boxed-in and even self-referential (e.g., Cole 1998), but professions also provide a platform for quality standards. Especially the smaller knowledge centres seem to rely blindly on the professional standards of their *suppliers* of knowledge. They stress the very low threshold access to easily digestible bits of information, whereby such complex problems of knowledge uncertainty or problem framing run the risk of being swept under the carpet. To be sure, there is attention for such problems in the larger knowledge centres, but most knowledge centres have to legitimate their existence by providing ready-made knowledge for policy, if need be at the expense of complication (Janssen and Schouw 2003). Knowledge centres may be useful for policy makers as a means to break through the iron triangles of corporatists structures, but they are also at risk of becoming an excessively instrumental and under-critical, but cheap resource for policy makers, which even lend themselves to token policy making (Beemer and Den Boer 2003).

Another striking development in the world of Dutch expertise for policy is an increased level of reflexivity among some of the major expert organisations. Major advisory organisations have produced reports touching on the status of knowledge in the policy process (in most cases their own knowledge), or have published (self-)evaluations. Reflection on policy research has a tradition in the Netherlands, especially as supported by the 'sector councils.' These councils of researchers, policy makers, and societal representatives traditionally reported on strategic goals for research in agriculture, health, nature and environment, and development. However, the focus on strategic research goals and recommendations on how to achieve them, has now been complemented with reflection on how this research is to relate to policy and how expertise is to be organised (Hoppe and Huijs 2003; Raad voor Ruimtelijk Milieu- en NatuurOnderzoek 2000; In't Veld 2000). In this new frame of thinking, the sector councils too see themselves increasingly as (also) knowledge brokers.

Several other examples stand out. The Scientific Council for Government Policy has produced several reports reflecting on expert policy advice, such as on uncertainty in environmental expertise (Wetenschappelijke Raad voor het Regeringsbeleid 1994), or recently on ICT and policy knowledge (Wetenschappelijke Raad voor het Regeringsbeleid 2002) or the public role of knowledge (Dijstelbloem and Schuyt 2002; Dijstelbloem and Schuyt 2003). Uncertainty in expertise for policy is a theme that is receiving increasing attention, especially in environmental issues. Since about 1997, uncertainty of expertise has become a topic at the RIVM, exacerbated by a media scandal in 1999 about the alleged over-reliance of the RIVM on computer models over actual measurements. Since then, RIVM has continued to organise (external) reflection of how it handles uncertainty and how it could construct better uncertainty management, some of which we are currently involved in (Van Asselt et al. 2001; Rijksinstituut voor Volksgezondheid en Milieu/Milieu- en NatuurPlanbureau 2003). Also noteworthy is the increased use of external evaluation reports as an occasion for active reflection on the operation of advisory institutes. Unlike a decade ago, evaluation reports are now available for several advisory bodies.

The increased reflexivity has come with a more relaxed attitude about allowing outsiders a glimpse into the back regions of expert knowledge production, glimpses which would have been considered inappropriate and undermining only ten years ago. However, the development is not shared everywhere. One reason is the lingering fear of making visible some of the contingent aspects of the construction of expertise, hence undermining credibility of expert advice. There are bureaucratic survival issues also. Even the suggestion of a negative evaluation report can have severe consequences for the continuation of advisory institutes, especially if their legitimacy was not entirely solid or in times of budget cuts.

A more ambivalent, but interesting development is the expansion of the expert resources of the Dutch Parliament. In Dutch government, the centre of gravity in expertise lies with the ministries. The most important among them have direct access to major research institutes and expert advisory boards, neatly organised around their respective jurisdictions. In contrast, the Parliamentary resources were traditionally limited to a small library staff and modest administrative support. Even the larger political parties can only support research bureaus with about half a dozen researchers. Expanding Parliamentary expert resources was seen as a wasteful duplication of governmental bureaucracy and a source of instability for the detailed political agreements of the executive that form the basis of coalition governments. As a rule, MPs hence have to rely on Ministers for information, through oral and written questions in Parliament, or through motions. To the extent that the traditions of Dutch Parliament provide research resources, they take the form of Parliamentary investigations, executed by ad-hoc Parliamentary committees; or budgetary oversight, supported by reports from the feared and very old Court of Audit (*Algemene Rekenkamer*).

In principle, Parliament is the focal point of public deliberation and hence a key place to look for deliberative expertise. Over the last decade, the position of Parliament in matters of expert knowledge has been reinforced. In existing institutions, there has been a more intensive use of Parliamentary investigations and an important shift in the Court of Audit from budgetary oversight to 'effectiveness' of policy – and hence substantive policy evaluation. In addition, some newer institutions have appeared on the Parliamentary horizon. During the first half of the nineties, the Rathenau Institute, the Dutch organisation for technology assessment, came to consider Parliament as its main client, a view that was formalised in a new legal mandate in 1994. Recently, the Rathenau Institute has supported some of the research activities of Parliament, for example by providing expertise for the organisation of Parliamentary hearings, but also for Parliament's new Research and Verification Bureau. This was installed in 2002 and is to support Parliament both with the 'verification' of expert reports offered by ministries, and with the commissioning of Parliamentary research, whether in the context of Parliamentary investigations or motions calling for research.

The list of expert organisations Parliament could consult is open-ended, as long as resources are available. Since the reorganisation of advisory councils of 1997, legal provisions were made for Parliament to ask any of the remaining advisory councils for advice directly. Remarkably, an exception is made for the planning bureaus. Without an official mandate, individual members of the planning bureaus are not even allowed to talk to MPs, as stipulated in the national civil service regulations. However, various Ministers have had their own views on the issue, sometimes even relaxing the reigns on direct Parliamentary contacts.[4]

So far, Parliament has made limited use of its new capacities to gather expert knowledge. With the support of the Research and Verification Bureau, eight research initiatives were undertaken since 2001, mostly through private consultants (Tweede Kamer 2002; Tweede Kamer 2003). In general, Parliament sticks to the more familiar instrument of Parliamentary investigation committees, especially for addressing problems perceived by a majority as particularly pressing. Its style of dealing with expertise so far has rarely followed the logic of deliberative expertise, where experts and policy makers communicate on a more equal footing. With the possible exception of a few parliamentary hearings and some minor experiments, the logic of representative government that makes use of instrumental expertise has dominated.

Here too, as with the other patterns, the pattern of deliberative expertise is therefore ambivalent and not without flaws. The new knowledge brokers in the knowledge centres may aim to improve public deliberation, but the risk of under-critical, unreflexive, and uncertified knowledge is clearly present. The reflection of large public expert organisations holds some potential for approaches to public deliberation of expert issues, but moves in this direction are very hesitant. Parliament seems to have an institutional gap to take a more active role in knowledge intensive policy issues, but leaves its instruments under-used or uses them in instrumental ways.

CHANGES IN THE POLITY, CHANGES IN SCIENTIFIC ADVICE

Changes in the expertise/policy making boundaries in the Netherlands over the last decade have been complex. New organisations, new formats for expertise, and new policy issues have emerged in new arrangements, while old ones have not necessarily disappeared. We have pointed at such salient developments as the restructuring of the advisory councils into a small set of expert councils; the gradual expansion of planning bureaus; the externalisation of departmental expertise; the contractualisation and commodification of expertise; the modest expansion of parliamentary expert resources; the growth of 'knowledge centres'; and the increasing reflexivity of expert organisations.

We have ordered these developments in three competing patterns for the organisation of the science/policy boundary. In corporatist patterns, where a limited set of actors is formally accredited to participate in decision making as representatives of societal interests, experts either participate to represent knowledge considered relevant in a corporatist style advocacy behind the scenes, or guard the boundaries of the playing field for corporatist negotiations. In the Netherlands, we have observed a shift from the first pattern to the latter. However, next to these, new patterns are becoming stronger. One is a neo-liberal pattern, which removes expertise from the state and its negotiation structures and uses the market as a means to coordinate expertise. The pattern does not come without a price. Transaction costs tend to be high, in the form of increasingly complex contractualisation of professional work, which tended to rely heavily on personal trust and negotiation. In addition, there are risks of undermining accumulation of knowledge over time, as past knowledge is stored in volatile consultancy firms and untraceable grey literature, and of instrumentalising research and hence undermining quality. Last, we see a number of these developments as examples of a deliberative pattern in the science/policy boundary, where public decision making is predominantly seen as a matter of collective reasoning and argumentation, stressing large degrees of participation in matters of interest as well as knowledge. In this pattern, the inclusion of the plurality of sources of knowledge is stressed, leading to new ways of integrating heterogeneous expertise, under increasing reflexive awareness of its limitations. There are various ways in which this pattern can be played out, for example ranging from stronger expert resources for parliamentary debate to policy sector knowledge brokers.

Similar shifts can be found in other countries. For example, in Germany an expansion of parliamentary expert resources has also been noted (see Brown et al., this volume), as well as a similarly hesitating acknowledgement of the plurality and distributed nature of knowledge (Heinrichs 2002). Points of comparison can also be found in other European countries (Glynn et al. 2001). Such shifts reshuffle the division of labour between experts and policy makers. For example, handling and interpreting uncertainties is a task for the experts where they are the linesmen of politics; but for the policy maker where expertise is hired on the market on an ad-hoc basis;

and one that is typically shared in collective reflection in deliberative patterns of organising expertise. Similarly, experts taking up tasks of conflict mediation is seen as perfectly legitimate in the deliberative pattern, but something to be avoided by the corporatist linesmen.

Our objective here is not to show how the Netherlands is unique or different from other countries, but to show that within a country such as the Netherlands, various patterns for organising science/policy boundaries are competing with each other. We have intentionally labelled these patterns with terms that allude to political connotations, both because of how the organisation of expertise is co-constructed with the organisation of political decision making, and because we want to point at the ideological connotations of these patterns. Rather than a grand transition from one mode of public expertise to another, or some essential national style, driven by a handful of constitutional prime movers, we see multiple patterns in tension and competition with each other. These patterns conflict and vie for dominance, argue against each other, and hence partly develop in response to each other. Such is the make-up of modern polities and the fact that we find similar tensions in the organisation of expertise, only shows how much expertise has become embedded in these polities.

University of Twente, The Netherlands

ACKNOWLEDGMENTS

The authors are grateful for information and suggestions from Arthur Petersen, Huub Dijstelbloem, W. van Honstede, Bert de Wit, Peter van der Knaap, Arno Korsten, Johan Melse, the participants in the 'Torentjesoverleg' (junior researcher seminar series of the 'Rethinking' programme), the editors of the Yearbook, and the participants in the preparatory workshop for this volume.

NOTES

[1] Although referred to as 'calculating through the proposals', this is more than just a matter of calculation, as negotiation and personalised expertise is required to interpret proposals and conform them to model input parameters.

[2] Known as 'Spatial Planning Bureau'. Official Dutch translations stubbornly use 'spatial' to refer to urban and regional planning.

[3] In 1988, there were 256 thousand 18-year olds, the age at which Dutch students normally enter higher education. Presently, the number is 184 thousand (Centraal Bureau voor de Statistiek 2003).

[4] One last addition is the recent announcement of the installation of a council of economic advisers. Three top economists are to provide Parliament with countervailing analytic power against the weight of the Central Planning Bureau. It is as yet entirely unclear how this council will operate, but previous experiences with a similar council in the UK are not very reassuring Collins, H. M. and Pinch, T. J. (1998), *The Golem at Large: What You Should Know About Science (2nd edition)*, Cambridge: Cambridge University Press.

SCIENCE/POLICY BOUNDARIES

REFERENCES

Abbott, A. (1988), *The System of Professions: An Essay on the Division of Expert Labour*, Chicago, Ill.: University of Chicago Press.

Asselt Van, M.B.A., R. Langendonck, F. v. Asten, A. v.d. Giessen, P.H.M. Janssen, P.S.C. Heuberger and I. Geuskens (2001), *Uncertainty and RIVM's Environmental Outlooks*, Bilthoven: RIVM, report 550002001.

Bakker, W. and F. Van Waarden (eds.), (1999), *Ruimte Rond Regels: Stijlen van Regulering en Beleidsuitvoering Vergeleken*, Amsterdam: Boom.

Bal, R., W. Bijker and R. Hendriks (2002), *Paradox van Wetenschappelijk Gezag: Over de Maatschappelijke Invloed van Adviezen van de Gezondheidsraad*, The Hague: Gezondheidsraad.

Beck, U. ([1986]1992), *The Risk Society: Towards a New Modernity*, London: Sage.

Beemer, F.A. and M.C. Den Boer (2003), 'Heeft de keizer kleren aan? Kenniscentra in het publieke domein', *Bestuurskunde* 12, 4: 170–7.

Beker, M., M. Ooijens and E. De Gier (2003), *Bewijs van Goed Beleid: Naar een Betere Verhouding Tussen Wetenschap en Sociaal Beleid in Nederland*, Amsterdam: SISWO, Working Paper.

Brickman, R., S. Jasanoff and T. Ilgen (1985), *Controlling Chemicals: The Politics of Regulating Chemicals in Europe and the United States*, Ithaca, NY: Cornell University Press.

Centraal Bureau voor de Statistiek (2003), *Kennis en Economie 2002: Onderzoek en Innovatie in Nederland*, Voorburg: Centraal Bureau voor de Statistiek, K-300.

Centraal Planbureau (2003a), *Charting Choices 2003-2006, Economic Effects of Eight Election Platforms*, The Hague: Centraal Planbureau, report no. 19.

Centraal Planbureau (2003b), *CPB website* [Internet] Available from: <http://www.cpb.nl> [Accessed October 2003].

Centraal Planbureau (2003c), *Through the Looking Glass: A Self-Assessment of CPB Netherlands Bureau for Economic Policy Analysis*, The Hague: CPB.

Cole, S.A. (1998), 'Witnessing identification: Latent fingerprinting evidence and expert knowledge', *Social Studies of Science* 28, 5: 687–713.

Collins, H.M. and T.J. Pinch (1998), *The Golem at Large: What You Should Know About Science (2nd edition)*, Cambridge: Cambridge University Press.

Berg Van den, H., G. Both and P. Basset (eds), (1993), *Het Centraal Planbureau in Politieke Zaken*, Amsterdam: Wetenschappelijke Bureau Groen Links.

Bogaard Van den, A. (1998), *Configuring the Economy: The Emergence of a Modelling Practice in the Netherlands, 1920–1955*, Doctoral dissertation, University of Amsterdam.

Dijstelbloem, H. and C.J.M. Schuyt (2002), *De Publieke Dimensie van Kennis*, Den Haag: Wetenschappelijke Raad voor het Regeringsbeleid, Sdu Uitgevers, Voorstudies en achtergronden V110.

Dijstelbloem, H. and C.J.M. Schuyt (2003), 'Kennismaken: Nieuw overheidsbeleid voor de kennissamenleving', *Bestuurskunde* 12, 4: 152–9.

Est Van, R., J. Van Eijndhoven, W. Aarts and A. Loeber (2002), 'The Netherlands: Seeking to involve wider publics in technology assessment', in S. Joss and S. Bellucci (eds.), *Participatory Technology Assessment: European Perspectives*, London: Centre for the Study of Democracy.

Expertisecentrum LNV (2003a), *Jaarverslag Expertisecentrum LNV*, Ede: Expertisecentrum LNV, EC-LNV nr. 2003/194.

Expertisecentrum LNV (2003b), *Website Expertisecentrum LNV* [Internet] Available from: <http://www.minlnv.nl/lnv/algemeen/eclnv/> [Accessed November 2003].

Gibbons, M. (1994), *The New Production of Knowledge: The Dynamics of Science and Research in Contemporary Societies*, London: Sage.

Gieryn, T. (1995), 'Boundaries of science', in S. Jasanoff, G.E. Markle, L.C. Petersen and T.J. Pinch (eds.), *Handbook of Science and Technology Studies*, Thousand Oaks: Sage, pp. 293–443.

Gieryn, T. (1999), *Cultural Boundaries of Science: Credibility on the Line*, Chicago: Chicago University Press.

Glynn, S., K. Flanagan, M. Keenan and D. Ibarreta (2001), *Science and Governance: Describing and Typifying the Scientific Advice Structure in the Policy Making Process – A Multi-National Study*, Seville: Joint Research Centre, Institute for Prospective Technological Studies, European Commission, EUR 19830 EN.

Hajer, M. and H. Wagenaar (2003), *Deliberative Policy Analysis: Understanding Governance in the Network Society*, Cambridge: Cambridge University Press.

Halffman, W. (2003), *Boundaries of Regulatory Science: Eco/toxicology and aquatic hazards of chemicals in the US, England, and the Netherlands, 1970–1995*, Doctoral thesis, University of Amsterdam.

Heinrichs, H. (2002), *Politikberatung in de Wissensgesellschaft: Eine Analyse umweltpolitischer Beratungssysteme*, Wiesbaden: Deutscher Universitäts-Verlag.

Hemerijck, A.C. (1994), 'Hardnekkigheid van corporatistisch beleid in Nederland', *Beleid & Maatschappij*, **21**, 1–2: 23–47.

Hoppe, R. (2002a), 'Rethinking the puzzles of the science-policy nexus: Boundary traffic, boundary work and the mutual transgression between STS and Policy Studies', Paper presented at *EASST 2002 Responsibility under Uncertainty*, York, 31 July – 3 August 2002a.

Hoppe, R. (2002b), *Van Flipperkast naar Grensverkeer: Veranderende Visies op de Relatie Tussen Wetenschap en Beleid*, The Hague: Adviesraad voor het Wetenschaps- en Technologiebeleid, AWT achtergrondstudie 25.

Hoppe, R. and Huijs, S. (2003), *Werk op de Grens Tussen Wetenschap en Beleid: Paradoxen en Dilemma's*, Den Haag: RMNO, RMNO publicatie 157.

In't Veld, R. (ed.), (2000), *Willens en Wetens*, Utrecht: RMNO/Lemma.

Janssen, J. and G. Schouw (2003), 'Verdiepen, verbinden, versterken: Het kenniscentrum grote steden in actie', *Bestuurskunde* **12**, 4: 178–89.

Jasanoff, S. (1990), *The Fifth Branch: Science Advisers as Policy Makers*, Cambridge, Mass.: Harvard University Press.

KEI Kenniscentrum Stedelijke Vernieuwing (2003), *KEI website* [Internet] Available from: <http://www.kei-centrum.nl/> [Accessed October 2003].

Kenniscentrum Grote Steden (2003), *Website Kenniscentrum Grote Steden* [Internet] Available from: <http://www.kenniscentrumgrotesteden.nl> [Accessed 22 October 2003].

Ketting, E. (2002), *Kenniscentra in Nederland*, Den Haag: Sociaal Cultureel Planbureau, SCP werkdocument 88.

Klink, E. (2000), *Pleitbezorgers en Policy Windows: De Institutionalisering van de Integratie van Emancipatie-Aspecten in het Nieuwe Adviesstelsel*, Doctoraal scriptie, Universiteit Leiden.

Köbben, A. and H. Tromp (1999), *De Onwelkome Boodschap: Of hoe de Vrijheid van Wetenschap Bedreigd Wordt*, Amsterdam: Jan Mets.

Kronje, G. (2003), *Overheid, Wetenschap en Toekomstverkenningen*, The Hague: Wetenschappelijke Raad voor het Regeringsbeleid, Werkdocumenten W130.

MacRae, D.J. and D. Whittington (1997), *Expert Advice for Policy Choice: Analysis and Discourse*, Washington DC: Georgetown University Press.

Ministerie van Landbouw, N.e.V. (2003), *Website Ministerie LNV* [Internet] Available from: <http://www.minlnv.nl> [Accessed November 2003].

Ministerie van Sociale Zaken en Werkgelegenheid (2003), *Website Ministerie van Sociale en Werkgelegenheid* [Internet] Available from: <http://www.minszw.nl> [Accessed November 2003].

Ministerie van Verkeer en Waterstaat (2003), *Website Verkeer en Waterstaat* [Internet] Available from: <http://www.verkeerenwaterstaat.nl> [Accessed November 2003].

Nationaal Dubo Centrum (2003), *Website Nationaal Dubo Centrum* [Internet] Available from: <http://www.dubo-centrum.nl> [Accessed October 2003].

Oldersma, G.J. (2002), 'Adviescolleges', in H. Daalder, R.A. Koole, J.W. Becker and F.M. v.d. Meer (eds.), *Compendium voor Politiek en Samenleving in Nederland*, Alphen a/d Rijn: Kluwer, p. C1100.

Overleg Directeuren Planbureaus (1996), *Protocol voor de Planbureaufunctie van CPB, RIVM, RPD en SCP* [Internet] Available from: <http://www.cpb.nl/nl/general/protocol.html> [Accessed October 2003].

Paardekooper, C. (2003), 'Kennisintensieve beleidsorganisaties', *Bestuurskunde* **12**, 4: 160–9.

Peters, B.G. and A. Barker (1993), *Advising West European Government: Inquiries, Expertise and Public Policy*, Edinburgh: Edinburgh University Press.

Raad voor Ruimtelijk Milieu- en NatuurOnderzoek (2000), *Andere Sturing: Andere Kennis Nodig?*, Den Haag: RMNO.

Ramdharie, S. and J. Trommelen (2003), 'Rapport kraakt relatie Suriname', *Volkskrant*, 7 November, p. 1.

Ravetz, J. (1999), 'What is post-normal science?', *Futures* **31**: 647–54.

Renn, O. (1995), 'Styles of using scientific expertise: A comparative framework', *Science and Public Policy* **22**, 3: 147–56.

Rijksinstituut voor Volksgezondheid en Milieu (2003), *RIVM website* [Internet] Available from: <http://www.rivm.nl> [Accessed October 2003].

Rijksinstituut voor Volksgezondheid en Milieu/Milieu- en NatuurPlanbureau (2003), *Leidraad voor Omgaan met Onzekerheden*, Bilthoven: RIVM/MNP.

Schreuder, A. (2003), 'Spreekverbod ecologen', *NRC Handelsblad*, 13 October, p. 3.

Shapin, S. (1992), 'Discipline and bounding: The history and sociology of science as seen through the externalism-internalism debate', *History of Science* **30**, 90: 333–69.

Sociaal Wetenschappelijke Raad (2000), 'Sociale wetenschappen en beleid: Een spannende verhouding', Paper presented at *Lustrumconferentie SWR 1995–1999*, Amsterdam.

Tweede Kamer (2002), *Raming voor de Tweede Kamer in 2003 Benodigde Uitgaven, Alsmede Aanwijzing en Raming Van Ontvangsten*, Draaiboek Onderzoek (Bijlage 3), Den Haag: Tweede Kamer, vergaderjaar 2001–2002, 28 336, nr. 8.

Tweede Kamer (2003), *Website Onderzoeks- en Verificatie Bureau* [Internet] Available from: <http://www.tweede-kamer.nl/organisatie/Onderzoeks_en_Verificatie_Bureau/index.jsp> [Accessed October 2003].

Vogel, D. (1986), *National Styles of Regulation: Environmental Policy in Great Britain and the United States*, Ithaca, NY: Cornell University Press.

Wetenschappelijke Raad voor het Regeringsbeleid (1977), *Overzicht Externe Adviesorganen van de Centrale Overheid*, Den Haag: WRR, Rapport Nr. 11.

Wetenschappelijke Raad voor het Regeringsbeleid (1994), *Duurzame Risico's: Een Blijvend Gegeven*, The Hague: Sdu Uitgevers, Rapporten aan de Regering Nr.44.

Wetenschappelijke Raad voor het Regeringsbeleid (2001), *Spiegel naar de Toekomst: Evaluatie van de WRR*, Den Haag: Wetenschappelijke Raad voor het Regeringsbeleid.

Wetenschappelijke Raad voor het Regeringsbeleid (2002), *Van Oude en Nieuwe Kennis: De Gevolgen van ICT voor het Kennisbeleid*, The Hague: Sdu Uitgevers, Rapporten aan de Regering Nr. 61.

Wit De, B. (2003), *Discoursen Achter Kennissystemen van Departementen*, memo, Den Haag.

CHAPTER 9

HEATHER DOUGLAS

INSERTING THE PUBLIC INTO SCIENCE

Over the past decade, attention has been increasingly focused on the problems of public participation in technical decision-making. The reasons for this attention are many: technical decision-making has become a locus of controversy in our political institutions; public dissatisfaction with these decisions seems only to rise; at the same time, experts continue to hold public opinions about these decisions in low regard. In the U.S., the Congress has attempted a legislative response to some of these issues, passing the Data Quality Act in 2001. In this act, the Office of Management and Budget is asked to ensure the integrity and objectivity of the science to be used in policy-making. It is doubtful that this effort will end the 'sound science'/'junk science' debates that have pervaded science-based policy-making. That the assurances of the Data Quality Act will quell public contention of policy-making appears even more doubtful. Yet scientists see little reason to think that increasing the involvement of the public in the development and evaluation of the science to be used in policy-making would improve the process and ease the debates. However, in this paper I will argue that, under some circumstances, it can do precisely that.

I am certainly not the first to offer the possibility of public involvement as a potential solution to debates around science in policy-making. Calls for increasing the quality and quantity of such involvement extend back at least 20 years. In the past decade, there has been an increase in the number of empirical studies of such processes, with attempts to determine what has been successful and what has not. Yet the literature seems to be plagued with two problems: 1) What evaluation structure should be used for these empirical studies is an open debate; and 2) Many authors still complain of the lack of empirical work in general. In this paper, I address both of these problems. First, I propose a basis for evaluating public participation in these processes, one that is grounded in a philosophical understanding of scientific knowledge and that aims to transcend the debates over which democratic ideal we should pursue. Second, I collect (and evaluate) empirical studies of public participation in technical decision-making from the past decade.

Why propose yet another normative measure with which to evaluate public participation processes? Several yardsticks for evaluating public participation have been proposed over the past decade. They include evaluating whether public participation has made a final decision more acceptable in practice (instrumental considerations), examining the substance of decisions to see whether more information is incorpo-

153

Sabine Maasen and Peter Weingart (eds.), Democratization of Expertise? Exploring Novel Forms of Scientific Advice in Political Decision-Making – Sociology of the Sciences, vol. 24, 153–169.
© Springer Science+Business Media B.V. 2009

rated into the decision-making (substantive issues), and considering whether citizen involvement has improved the democratic legitimacy of a decision (normative issues) (Fiorino 1990; Laird 1993). In practice, these yardsticks often translate into functional considerations such as: evaluating the process by which the public is involved for attributes of fairness, determining whether the public has any actual impact on the decision, and whether the public (and sometimes the experts) learn anything in the process (Rowe and Frewer 2000; Renn et al. 1994). The normative rationales for these various yardsticks generally center on ideals of democracy. However, there are competing ideals in the literature as to what democracy should entail (Laird 1993; Fischer 1993). For example, depending on whether one ascribes to a 'direct' democracy ideal or a 'liberal' democracy ideal, very different standards for adequate public participation arise. In addition, neither ideal provides a clear rationale for why the public should be involved not just with the policy decisions, but also with the performance and evaluation of scientific studies on which the policy is to be based. Yet so often, policy disputes center on whether or not public actors accept or reject the scientific basis for policy-making.

Working from a philosophy of science perspective, I articulate a rationale for public participation in the development and interpretation of science to be used in policy-making. Because ethical values are needed in the practices of science throughout the research process, some accountability for those values is also needed. Regardless of which democratic ideal one holds, the values used to do scientific analyses that then inform public policy should reflect public values. Different processes can then be evaluated by the extent to which they allow citizens to inform the values used in doing the relevant technical analyses, and I examine several prominent and promising mechanisms for achieving a productive interaction between experts and the public.[1] In essence I ask: Has citizen involvement helped to bring citizen values into the heart of technical judgment? The extent to which this can be achieved undergirds the instrumental, substantive, and normative yardsticks mentioned above.

THE PHILOSOPHICAL PERSPECTIVE: VALUES IN SCIENCE AND POLICY-MAKING

The role of values in science has been a source of steady controversy in the philosophy of science over the past fifty years. The standard position has settled into the following: while values invariably creep into science (because scientists are human), scientists should make every effort to limit values to the external aspects of science (choice of research problems, application of science and technology). The exception to this normative rule is that epistemic values (i.e., concern with empirical accuracy, scope, simplicity, fruitfulness, internal consistency, and explanatory power) are acceptable throughout science. (Kuhn (1977) provides a classic account of epistemic values; Lacey (1999) attempts a defense of the standard position.) This norm prevails in realms beyond the esoteric halls of philosophy of science. It is this norm that underpins the widely held view of science as a value-free affair, that supports particular pedagogical approaches to teaching science (the 'answers in the back' approach), and that in part undergirds efforts to clearly define a realm for science distinct from the realm of policy-making, where values are recognized as necessarily pervasive. Under this view, the belief is that if only we can more carefully construct the border between science and policy, we can reaffirm the value-free nature of science. (Rosenstock and Lee (2002) provide a similar argument; Douglas (2004b) argues against having a value-based border between science and policy.)

As I have argued elsewhere, however, the norm of value-free science is a bad norm (Douglas 2000). Scientists must make a series of choices throughout the scientific process. Once they have framed a problem, they must decide upon a methodology to address it. When collecting data, they must decide how to record unexpected or borderline results. They also must decide when to reject data as unreliable because of some uncontrolled factor in the experimental process. They must then decide how to interpret their data. Only then can the results be used in a policy-making process or applied in some context. While scientists may hope for little need of judgment in their choices, disagreements among scientists concerning the appropriate methodologies, the quality of data, and the correct interpretations of data belie the need for such judgments. Clearly there are differences of expert opinion on how to best perform studies, particularly in developing areas of research or on the 'cutting edge.' Where such differences exist, judgment is needed. Because much of the science needed to make policy falls into these developing areas of science rife with contention among scientists, such science is also rife with the need for scientists' judgments.

How should scientists make these judgments? As noted above, the standard answer in philosophy of science is that scientists should consider only 'epistemic' values.[2] However, this answer is based on the assumption of scientific isolationism, i.e., that scientists operate in an enclave that is largely separate from society at large, taking in resources from society and, when scientific consensus is achieved, revealing answers to society. The actual practice of scientists in advisory roles in the past fifty years and the importance of tentative results in shaping public debate are in direct contradiction with this view. If we reject scientific isolationism for the fiction it is, there is no reason to restrict the basis for scientific judgment to 'epistemic' values only. Indeed there is good reason to require the consideration of ethical values throughout the scientific process. Although it may go against the current norms of scientific practice, I have argued (Douglas 2000) that if we are to hold scientists to the same moral norms as the rest of us, scientists must consider ethical consequences of error in their work. And I have also argued that we should hold scientists to the same moral norms as the rest of us (Douglas 2002). Thus, in policy contexts with uncertain science, scientists must use ethical values in their work.

To see this necessity, consider a situation in which there are extra-scientific consequences of error. If a scientist makes a judgment and gets it wrong, who (outside of the scientific community) gets hurt? And if he makes a different judgment and gets it wrong, who else is harmed? The values one places on the costs of error beyond science are ethical values. Because we all share the moral responsibility to consider potential consequences of error in our daily lives (Douglas 2002), scientists need to use ethical values to determine which errors in their work they should be more careful to avoid (e.g., false positives or false negatives). Only if one thinks there are no consequences of error in science beyond the enclave of science can one suggest that scientists need not weigh those consequences in making choices. Because that position is clearly false (particularly in science used for policy-making), we are forced to the view that scientists *must* use ethical values in the process of making judgments while doing science. If scientific methodologies improve and predictive accuracy increases (a boon to policy-makers hoping for effective actions), scientists may need to worry less about potential consequences of error as the chance of error decreases.[3] But because some uncertainty is ineliminable, there is no complete removal of this need. When there are choices to be made in science and there are ethical consequences of error, there must be ethical values.

Whose values should they be? This question raises the specter of democratic concerns. Regardless of which theoretical ideal of democracy one might hold, it is not acceptable for a minority elite to impose their values on the general populace. If scientists can make these judgments in private, not disclosing them in their published work, and thus shape public policy through these judgments with no possible avenue for public accountability, any standard of democracy will have been violated. Thus minimally, scientists need to be more explicit in their work concerning where judgments are made and how they made them, including a discussion of values used. While many might think such behavior would threaten 'scientific objectivity,' I have argued elsewhere that this need not be the case (Douglas 2004a). Clear discussion of legitimate value judgments, e.g., those that are needed to weigh the consequences of error among multiple sound methodologies, need have no harmful repercussions for objectivity *per se*. The problem for values and objectivity arises, rather, when values take the place of evidence, or when values lead one to simply ignore evidence that runs contrary to a desired outcome. Proper and necessary consideration of ethical values in places of needed scientific judgment pose no threat to objectivity as such.

While disclosing the judgments made in science and the values used to make those judgments is a good first step, that does not resolve the problem of whose values should guide judgments made in science and in science-based policy-making. An ideal situation would be to have a public debate over contested values, resolve the debate, and then ensure that scientists employ those values when making their judgments in practice. However, many of the value disputes have only recently surfaced and are just becoming defined, much less resolved. For example, debate continues to rage over what our obligations are to future generations as opposed to current ones in less-developed areas, whether we have rights to be free of health risks or whether some risks can be imposed on all for the greater good, whether gaining some degree of economic benefit is worth losing some degree of health for humans or ecosystems, and further which is worse for human health: reduced wealth or increased chemical exposure. While we may hold out hope that some good public debates guided by sound ethical argumentation will help resolve these disputes, or at least narrow the range of plausible positions, we should not wait for this outcome. In the meantime, we can develop better processes, ones that allow citizens to help direct the science used to make policy and to help interpret that science for policy, and ones that allow scientists to better understand the value concerns of citizens. Developing these processes may in the end also promote the needed ethical debates.

The need for improved processes has been widely recognized. Grass-roots calls for more public involvement and greater public control of technical decision-making are prolific (for example, O'Connor 1993). Political theorists (as noted above) have provided a range of reasons for increased involvement. Most surprising, however, are calls for improved processes from bastions of science. A prominent example can be found in the 1996 U.S. National Research Council's (NRC) *Understanding Risk* (also known as the 'orangebook'), which redefined the risk analysis process through its discussion of risk characterization. This example is surprising in part because an earlier NRC report, *Risk Assessment in the Federal Government,* the canonical 1983 'redbook,' argues for a strict separation (as much as possible) between the expert work of risk assessment and the citizenry involvement with risk management. In the redbook, the NRC attempted to conceptually distinguish and to practically differentiate the risk assessment process (which was to be as scientific and value-free as possible) from the policy-laden and value-laden risk management process. The point of risk characterization in the redbook was simply one of summarizing the scientific results of risk assessment into a useable form for risk management (NRC 1983: 20; Stern and Fineberg 1996: 14).

When the 1996 NRC panel was asked to provide a closer examination of risk characterization, the authors redefined risk characterization from a brief transition between risk assessment and risk management to a process that should "determine the scope and nature of risk analysis" (Stern and Fineberg 1996: 2). Risk characterization became the framework for the entire risk study and decision-making process. The NRC panel then defined risk characterization as an 'analytic-deliberative process,' with potential roles for both citizens and scientific experts. Analysis is defined as the use of "rigorous, replicable methods, evaluated under the agreed protocols of an expert community." Deliberation is defined as "any formal or informal process for communication and collective consideration of issues" (Stern and Fineberg 1996: 4). While these processes are defined as distinct, they are conceived by the NRC as being in a continual state of interaction throughout the assessment and regulatory process. Both are always needed: "deliberation frames analysis, and analysis informs deliberation" (Stern and Fineberg 1996: 30). Rather than ghettoizing the public to the end stage of decision-making, the NRC now appeared to provide support for the public's involvement throughout the process. What this looks like in practice can now be explored.

PUBLIC PARTICIPATION IN SCIENCE AND POLICY-MAKING

As noted above, there have been increasing numbers of empirical studies focused on public participation in the past decade, with an accompanying proliferation of potential techniques for evaluating the success of these various mechanisms. Given the theoretical considerations I mention above, I propose another possible evaluation measure: the extent to which a process maximizes the interaction between citizens and experts, and maximizes the influence they have on each other. If deliberation is truly needed to inform analysis, and analysis to inform deliberation, experts and citizens need to be working in close contact to address our most difficult science-based policy questions.

To understand what this would mean in practice, consider the standard model in which experts and the public rarely interact. On the one hand, we have those processes in which citizens have a great deal of interaction amongst themselves, with strong deliberative processes, but little interaction with or impact on experts. Experts may be brought in as a source of information, but are not expected to take away from that experience anything important for their own work. Examples of this may include (depending on the details of the process) citizen consensus conferences or citizen panels where the expert analytical work is already complete. In these processes, there is a good range of public views and values being discussed, but little influence on experts. On the other hand, we have those processes composed primarily of expert work, complex analyses performed without public input, that are then placed before the public for formal approval and use. Citizens perform little deliberation and the process is dominated by analysis and deliberation among experts only. There is little assurance that an appropriate set of values has informed the expert work. Examples here may include expert panels accompanied by public hearings or referenda. With both these types of processes, the analyses (and/or expert deliberation) are performed separately from the public.

What is needed, given the theoretical concerns articulated above, is to integrate these types of processes, to maximize citizen-expert interaction. If citizens were more fully involved with expert deliberations, the public could be assured that the values used to shape the technical analyses would be appropriate ones. For example, citizen panels could assist with the direction of scientific studies conducted to inform policy-making. The analytic and deliberative elements would be interconnected.

Getting the public involved directly in the study of technical issues by having them assist in guiding research has been called participatory research or collaborative analysis (or some similar combination). The theoretical advantage of such 'analytic-deliberative' processes has been described above. There are also practical advantages. When citizens provide input at the stage of regulatory decision-making, it is unclear whether such participation can move beyond what Boiko et al. calls the 'tokenism' level of citizen participation to the level of 'citizen power' (Boiko et al. 1996: 247). Beholden to legislatures, regulatory authorities cannot really share decision-making power with citizen *or* expert groups. However, that is true only for the final regulatory decisions made. Where citizens can have direct power (and where experts already have direct power) is in the technical studies and analyses that are performed to inform and support a regulatory decision. Citizen input here can have binding authority, if the study designers and officials allow this to take place.

Here I give a brief survey of three examples from the past decade in which citizens have been given the opportunity to direct technical analyses to be used in regulatory decision-making. As we will see, there seem to be three distinct ways in which citizen input to technical assessments and analyses can be valuable: 1) Citizens can help to better frame the problem to be addressed. (Are the appropriate range of issues and potential solutions being considered? Is the scope of the analysis appropriate?) 2) Citizens can help provide key knowledge of local conditions and practices relevant to the analyses. 3) Citizens can provide insight into the values that should shape the analyses. (How do citizens weigh the potential consequences of error? What kinds of uncertainties are acceptable or unacceptable? What assumptions should be used to structure the analyses?) This last point of input is both crucial and often overlooked. Because values are needed to shape analyses, whose values is important. Traditionally, the values have been both hidden and those of the experts making the judgments. Many experts think that citizens are unable to understand the technical complexities of analyses, much less provide guidance at points of expert judgment. Yet

INSERTING THE PUBLIC INTO SCIENCE 159

the examples below suggest ways in which citizens can do precisely that, with the result that experts think the analyses are strengthened *and* the citizenry trusts the study's results.[4]

ANALYTIC-DELIBERATIVE PROCESSES: COLLABORATIVE ANALYSIS IN PRACTICE

In this section, I will discuss three examples in which a collaborative analysis approach has been attempted. Not all of them were equally successful. The first is focused at the local level, with very successful results. The second is focused at a national level, with also successful results. Finally, I discuss a local attempt with less successful results, but which illustrates the need for care in the construction of the processes of collaborative analyses.

Valdez, Alaska and the Marine Oil Trade

In his detailed study of disputes concerning the marine oil trade in Valdez, Alaska, Busenberg compares a period in the early 1990s characterized by 'adversarial analysis' (i.e., competing experts utilized in disputes marked by a lack of trust and that are generally unsolvable) with a later period characterized by 'collaborative analysis' (Busenberg 1999). Drawing from Ozawa (1991) for his account of these two forms of interaction over policy disputes, Busenberg's account of collaborative analysis in Valdez is both intriguing and promising. The two opposing groups, a community group called the Regional Citizen's Advisory Council or RCAC (formed in 1989 after the Exxon Valdez oil spill) and the oil industry, had a history of distrustful and confrontational relations. By 1995 they both seemed to realize the impasse to which this generally led, and resolved to find a way around these difficulties. The dispute at that time centered around what kinds of tug vessels should be deployed in the Prince Williams Sound to help prevent oil spills. Instead of doing competing risk assessments to influence the policy decision, the RCAC, the oil industry, and the relevant government agencies decided to jointly sponsor and guide the needed risk assessment (Busenberg 1999: 6).

The risk assessment proceeded with a research team drawn from both industry and RCAC experts. The project was funded by the oil industry, RCAC, and the government agencies. The steering committee had representatives from all of these groups, and met regularly (fifteen times) to direct the study. Not surprisingly, members of the steering committee learned much about the intricacies of maritime risk assessment. More surprisingly, the research team found the guidance of the steering committee to be very helpful. As one quoted in Busenberg (1999) noted, the process "increased our understanding of the problem domain, and enabled us to get lots of data we didn't think was available ... the assumptions were brought out in painful detail and explained" (Busenberg 1999: 7f.). The additional data was needed when the committee decided that "existing maritime records were an insufficient source of data for the risk assessment models" (Busenberg 1999: 8). The steering committee then assisted the researchers in gaining more detailed data needed to do an adequate risk assessment. In sum, Busenberg's discussion suggests that all three of the ways in which citizen input can assist with a technical study were met: 1) the scope of the problem was better defined, 2) data quality was increased, and 3) assumptions and uncertainties were properly examined and weighed. The final risk assessment was accepted as authoritative by all parties, and one of the new tug vessels was deployed in the Sound in 1997 as a result (Busenberg 1999).

This example shows the promise of analytic-deliberative techniques when deployed at a local level to address local environmental issues. Where there are clearly defined stakeholders (both with some resources), they can decide to collaboratively design, fund, and direct research that can help resolve technically-based disputes. Crucial to the process is an equal sharing of power among the parties, made possible in part by the joint funding of the research. The public values find voice in the directing of the research, by helping to critically examine and shape the scope and guiding assumptions of the analysis. The next example examines whether this is possible at the national level.

Chemical Weapons Disposal Methods

As Futrell describes in his 2003 paper, how to dispose of chemical weapons has engendered a debate in the U.S. at both national and local levels. Throughout the 1980s, the controversy between the army and local citizen groups intensified as citizens became disenchanted with the army's perfunctory attempts at citizen input (Futrell 2003: 459–464). Citizens felt that key decisions, such as whether alternatives to weapons incineration would be seriously considered, were made prior to any opportunity for their input. By the late 1980s, citizens at eight disposal sites around the U.S. had banded together to form the Chemical Weapons Working Group. They pressed their case for serious consideration of alternatives to incineration, and succeeded in gaining national attention for the issue in the early 1990s (Futrell 2003: 465). By 1997, Congress ordered the Department of Defense to establish a new effort separate from the army's incineration program, called the Assembled Chemical Weapons Assessment (ACWA) program. The ACWA, with its independence from the incineration program, developed what Futrell calls a "participatory approach to decision making" (Futrell 2003: 466). The ACWA brought in an independent (private sector) mediator to manage a "Dialogue on Assembled Chemical Weapons Assessment" (Futrell 2003: 466). This Dialogue process consisted of a series of meetings at multiple sites around the country involving the local citizens, state regulators, national activists, and relevant federal officials. As Futrell describes it: "In face-to-face consensus meetings, Dialogue participants developed criteria against which alternative technologies would be assessed and provided input into each stage of the actual technical assessments" (Futrell 2003: 466). These criteria, Futrell notes, reflected the concerns that had been raised by citizens since the beginning of the controversy. The alternative technologies that came out of the ACWA are now under further evaluation and may be instituted at several sites (Futrell 2003: 468).

As with the Valdez example, the collaborative approach allowed for citizen values to be reflected in the technical analyses that followed. The Dialogue process enabled citizens to influence the criteria under which various technologies would be judged, and these criteria reflected citizen concerns and values. The collaborative approach also altered the framing of the issue, away from an 'incineration only' approach. It is less clear, however, that citizen input assisted with the collection of local data, although many early concerns of citizens focused on a lack of site specific risk assessment by the army. Even if all three benefits of citizen participation are not met in this example, the process did produce a "mutually acceptable study of chemical weapons disposal aimed at technically safer and more politically legitimate decisions" (Futrell 2003: 472). Whether the study's suggestions are fully utilized remains to be seen.

In my last example, a history of mistrust overwhelms the attempt at collaborative analysis. The analytic-deliberative process falls apart, in large measure because of process failures and a lack of commitment by the lead government agency.

Hanford, Washington and the Legacy of Plutonium Production

Hanford, Washington, became the first plutonium production site in the world in 1944. Plutonium production ended at Hanford in the 1980s, but it left Hanford one of the most contaminated sites in the United States. Because of its mission (the production of weapon-grade plutonium), a shroud of secrecy had always been drawn over the activities of Hanford. When citizens pushed for greater access to information about the site in the 1980s, they were appalled by what they found in released documents. From radioactive wastes stored in aging leaking tanks to intentional releases of radioactive iodine (the 'Green Run' of 1948), the actual state of the site was far different from the reassuring press releases the public had been fed for forty years. (Gerber (1997) provides a complete account of Hanford's history.) Area citizens have been involved with health assessment and clean-up efforts since the late 1980s, and the Department of Energy has been attempting to rebuild trust with these citizens (Kaplan 2000). Unfortunately, efforts have not gone well.

For example, one area of concern involves the contamination of groundwater at Hanford and the potential for subsequent contamination of the Columbia River. To help address this concern, the Department of Energy (DOE) initiated an assessment of the risks, the Columbia River Comprehensive Impact Assessment (CRCIA) (Kinney and Leschine 2002). As described in an analysis of the CRCIA process, it "was not a typical technical analysis, however. Tribal and stakeholder representatives had seats on the CRCIA Project Management Team and, in the parlance of NRC's analytic-deliberative process, participated as 'equally valid contributors'" (Kinney and Leschine 2002: 87). Unfortunately, Kinney and Leschine describe a process that falls apart as the citizen stakeholders attempt to shape the risk assessment. As the stakeholders pushed for a more comprehensive assessment that would more successfully address their concerns, the DOE split CRCIA into two parts, one to design and conduct a 'screening assessment' which would determine where the most serious risks would lie, and the other to design the more comprehensive assessment sought by the stakeholders. While the DOE contractor was conducting the first part, with full support from the DOE, stakeholders were investing their energies in the second part, which the DOE eventually repudiated.

The DOE's gradual withdrawal from and rejection of CRCIA Part II was unfortunate, because it would have provided an excellent example of collaborative analysis at work. The stakeholders working on Part II designed a process which would have allowed a high level of public involvement in the assessment process, with heavy emphasis on 'predecisional participation.' Kinney and Leshine describe this further:

> The type of predecisional participation the authors speak of is an oversight role in the day-to-day work of conducting risk assessments. Were their ideas to be implemented by the DOE, the management of risk assessment work would be carried out by a board composed of representatives from the socioeconomic groups who are affected by Hanford's clean-up and disposal decisions. This citizen management board, called the CRCIA Board, was envisioned to eliminate the need to make 'arbitrary assumptions' during the course of an assessment, as CRCIA Board approval would be required before any assumptions were incorporated into an analysis. In addition, the CRCIA Board would develop its own standards for data quality and maintain final authority over decisions relating to assessment protocols (Kinney and Leschine 2002: 89).

What had been proposed would have been a quintessential collaborative analysis. Recognizing the need to make choices and assumptions in the course of a technical assessment, the stakeholders wanted to have a say in those choices, to be able to shape those choices with their values. Unfortunately, the DOE has not accepted this plan. Indeed, the whole CRCIA process was fraught with problems. The DOE never clearly articulated its goals for the CRCIA Project Management Team (Kinney and Leschine 2002: 88). There were no clear process rules for the Team's meetings nor were any formal mechanisms used for dispute resolutions, and as a result, discussions among members were hampered (Kinney and Leschine 2002: 95). Finally, the DOE did not fully support the CRCIA process and was free to reject any outcomes, which they did. All of these factors led to a feeling of frustration and dissatisfaction among participants. An opportunity for collaborative analysis, and for a rebuilding of trust in the DOE, was lost.

Neighbors of Collaborative Analysis

The preceding examples are ones that attempted to maximize the public-expert interaction, and thus suggest that involving the public in science done for policy can work in practice. In the (successful) examples discussed above, citizens worked closely with experts to design and run a scientific or technical analysis relevant to policymaking. Other models of public participation with technical decision-making focus less on expert-public interaction. These 'neighbors' of collaborative analysis help to highlight the strengths and limitations of collaborative analysis.

Science Shops

Begun in the 1970s in the Netherlands, the idea of 'science shops,' an outreach institution within a university that makes its scientific expertise available to the public, has spread beyond its country of origin. Both Irwin (1995: 156) and Sclove (1995: 226) mention the growth of science shops in Western Europe, and both authors hold out science shops as a potentially promising technique for getting citizens better access to scientific expertise. A more recent analysis by Wachelder (2003) suggests difficulties with the science shop program as a result of social and economic changes in the Netherlands. More telling for my purposes, however, are the issues raised by Irwin in his earlier discussion of science shops. Irwin notes the inherent difficulties (aside from institutional resource constraints) that one should expect with science shops. He brings attention to the problem of whether citizens' questions and inquiries can be translated into a form that scientists find recognizable, and whether scientists' results can be effectively utilized by citizens. He notes, for example, that science shops seems to have had little effect on the general research agendas of scientists, although some commentators still think there is a good possibility for this to occur (Irwin 1995: 159; Sclove 1995: 226). More problematic are the issues arising in the framing of research, which Irwin sees as having been hampered by "the impossibility of achieving a workable dialogue" between citizens and scientists (Irwin 1995: 161). Yet Irwin also describes an example where this problem appears to have been surmounted, from the Northern Ireland Science Shop. In one case, citizens seem to have taken an active role in helping to determine the methodologies employed by the students at the science shop in researching a question using a survey. As Irwin describes:

> The survey was conducted with the assistance of a group of university students. However, rather than following the usual academic model, the survey contents were very much the subject of negotiation between community representatives and the students. We see here the emergence of a new style of 'scientific' inquiry—one which attempts to negotiate with the concerns and problem definitions of the concerned groups (Irwin 1995: 164).

As Irwin's discussion shows, this example verges on the model of collaborative analysis. In addition, it is clear that this is not the usual model of operations for science shops, and Irwin sees it as being relatively new and untested. If science shops are to maximize the citizen-expert interaction (and thus the reflection of citizen values in the science), this kind of collaboration and negotiation would have to become the norm.

Yet allowing this kind of negotiation between clients and scientists presents additional problems. Because the science shops are petitioned by particular groups (increasingly, industry or commercial groups according to Wachelder (2003)), the values shaping the studies would not be representative of citizens as a whole. With the collaborative analysis examples above, an array of stakeholders (and value positions) were involved with shaping studies. Such would not be the case in science shops, and thus allowing the client to shape the study could lead to charges of cooptation and a decrease in the credibility of the science shop. It seems that the strength of science shops lies in their ability to provide access to traditional scientific results with traditional scientific legitimacy that can then be used in political arenas. Whether we would want to perpetuate that purpose given the theoretical considerations discussed above remains an open question.

Citizen Planning Efforts: Locally Focused Discourse

In many local contexts, there has been an increased effort to involve the citizenry more directly in planning efforts. In this section I will discuss two recent examples, both of which provide clear promise for citizen involvement. The examples, however, provide very different models for the kind of interaction one might want between experts and the public. In the first example, Renn's 'cooperative discourse,' there is no *direct* interaction between expert groups and the public (Renn 1999). In Fischer's example from Kerala, India, there is no expert knowledge until the communities, in cooperation with experts, generate it (Fischer 2000: 157–167).

Renn's model for *cooperative discourse* divides the decision-making process into three parts: 1) A consultation with stakeholder groups to determine general concerns with an issue and the possible values at stake, resulting in a 'value-tree'; 2) A consultation with experts to rate various policy options and their consequences with respect to the values in the value-tree;[5] 3) A discourse aimed at choosing a policy option generated among randomly selected local citizens using the results from the expert consultation (Renn 1999: 3050f.). In this highly structured model, stakeholder values feed into expert views, which then help inform the final selection of policy options accomplished in the discourse among selected citizens. Renn describes several contexts in which this model has been used, and notes its successes (particularly in the European context). The model seems to be able to overcome some of the most stubborn problems of local planning and siting issues. As Renn noted in one case, the participants overcame the NIMBY syndrome: "The most outstanding result was that panelists were even willing to approve a siting decision that would affect their own community" (Renn 1999: 3052). In another case, Renn noted that complete agreement was achieved: "The most remarkable outcome was that each panel reached a unanimous decision" (Renn 1999: 3052). Thus, this method can produce a high degree of consensus among the citizen groups for particular policy options, overcoming even the resistance to siting undesirable facilities in one's own community. However, it does not allow for direct interaction between experts and the public, thus preventing local knowledge from playing a role in expert analyses. Nor is it clear to what extent expert analyses are actually shaped by public values. Whether new analyses are done in the light of public concerns is not mentioned. These differences highlight how cooperative discourse diverges from collaborative analysis in practice. Finally, citizen recommendations in these cases are for actual policy choices. Because the citizens are not elected officials (or accountable to elected officials), these choices are non-binding – and in some of the cases they are ignored (Renn 1999: 3052). Despite these difficulties, the structured approach of cooperative discourse may be particularly useful for some contexts in which direct expert-public interaction is not feasible.

Fischer's discussion of Kerala's planning efforts provides a very different model for how citizens and experts can work towards well-informed decisions, one that is much closer to collaborative analysis. Among the poorest of states in India, Kerala is an area that has a tradition of left-wing politics (Fischer 2000: 158). The local leadership decided that a general planning effort would be the best way to improve the lives of the population while involving the local citizenry in governmental decision-making. In line with this view, the State Planning Board devolved over a third of the planning efforts (and resources for those efforts) to the local level (Fischer 2000: 159). However, much of the local information needed for good planning (e.g., current land uses, soil types, etc.) had never been gathered. Thus began a genuine effort of 'participatory research' to produce resource maps that could then inform planning efforts (Fischer 2000: 161). Local volunteers were trained by appropriate experts to gather the requisite information (Fischer 2000: 165). By the end of that work, sets of detailed maps that could serve as a well-informed basis for planning discussion and decision-making were in hand. With these maps, local communities could develop 'action plans' that would encompass both the current state of the community and what the community wanted to change. Note how much closer this model comes to collaborative analysis, with experts working in tandem with local citizens to generate needed knowledge. (Because the amount of expert analysis needed in this example is not clear, I have not considered it a definitive example of collaborative analysis.)

In sum, discourses locally focused on planning efforts can take on some of the characteristics of collaborative analysis. They can draw extensively from local knowledge, while generating clear discussions of local needs and values. What is unclear in these examples is the degree to which citizens have the ability to direct experts in the performance of analyses useful to their deliberations. To the extent that citizens have this capacity, the process becomes an example of collaborative analysis.

Consensus Conferences: Discourse Beyond the Local Context

Consensus conferences have come to the fore in the past decade as a promising technique for increasing citizen input into complex policy decisions. Developed most thoroughly in Denmark, consensus conferences recruit average citizens for an intensive experience of education in an issue, group deliberations, further questioning of expert panels, and ultimately a drafting of a consensus statement on the issue (Joss and Durant 1995). The strengths of the approach are well documented: citizens involved learn a great deal about an issue; the group deliberation is often revealing of deep issues; and the consensus document often reflects well what the citizenry as a whole would think after similar depth of exposure (Sclove 2000). In the Danish context, where the consensus conferences are sponsored by the government, Joss (1998) has shown that they have had a positive impact on policy-makers (most notably legislators). In the American context, Guston's analysis (1999) suggests that the policy impact was negligible. This is not surprising given that the U.S. government has had little involvement with sponsoring or promoting consensus conferences.

Consensus conferences thus provide an interesting and potentially valuable technique for fostering careful deliberation concerning science and technology-based policy issues. However, it must be noted that consensus conferences do not currently provide for a way in which citizen deliberation can actually direct or shape analyses central to these issues. Citizens use already completed analytical work (presented to them by experts) to deliberate on the contentious issue. In the standard model, they do not direct any new work by the experts they question; the experts are there solely to help answer citizen questions.

This lack of mutual influence may suggest a new direction for consensus conferences. Because citizens are drawn from a random sample (or drawn randomly from interested citizens), no decision-making authority can be legally given to consensus conferences addressing policy choices. The 'moral authority' of the consensus conference can have a substantial political influence (Joss 1998), or not (Guston 1999).

But the power of consensus conferences to influence policy directly remains weak. However, consensus conferences that address value issues that arise in technical analyses could have a strong influence on later analyses. A general consensus on how to make trade-offs between economic gains and human health risks, for example, could be used to determine the statistical strength needed to say a correlation is 'significant.' The difficulty with using consensus conferences to reflect on the assumptions needed in analyses is that, depending on the context of a particular analysis, different assumptions may be warranted. How much a general consensus on a general set of issues would be of use to specific analyses that need to be performed remains to be seen.

The strength of consensus conferences may lay with the general level at which they function. By addressing fairly broad policy issues (e.g., genetic engineering, communications policy, global climate change), the consensus panels have the potential to broaden the scope of public debate generally, and may bring to light concerns experts and policy-makers had not previously considered. This interaction is not close enough to be called collaborative analysis, but it is highly beneficial nevertheless.

In sum, the three near neighbors of collaborative analysis I have discussed, science shops, local planning efforts, and consensus conferences, all exhibit some of the traits of analytic-deliberative processes. Yet, as I have noted, the interaction between experts and the public is not as intensive as it is with collaborative analysis. In general, the influence of the public on the experts remains muted. This need not be the case (as noted in the examples that verge on being collaborative analyses), but to alter that aspect of these processes may damage some of their other strengths. Whether the strengths of collaborative analysis are sufficient to counter potential losses remains to be determined.

CONCLUSION: STRENGTHS AND LIMITATIONS OF ANALYTIC-DELIBERATIVE PROCESSES

Following the theoretical considerations presented in the first section, values are needed to inform scientific and technical analyses. Yet the need for value judgments in these analyses may not be obvious at the start of an investigative process; even less obvious is the nature of concerns that might arise and which values should be used to make key judgments. An interactive process in which a range of citizens can help guide analyses as they move forward, an analytic-deliberative process, seems an ideal solution to this problem. Collaborative analysis provides a workable model of this process. In the (successful) examples discussed, citizens were able to inform the scope of the analyses, provide local knowledge to improve data quality, and ensure that appropriate values shaped important judgments. Because citizen involvement was prior to policy decision-making, their input on these points could be binding. Thus collaborative analysis has the potential to short-circuit many of the chronic problems in policy-making, including lack of public trust in technical work, lack of empowerment of citizens, and access to reliable data.

Yet problems remain for analytic-deliberative approaches. For example, one of the issues I have not discussed above is the selection of participants for analytic-deliberative processes. In my examples, the citizens involved with the process are those who recognize that they have an interest in the issue. The citizens involved with the Valdez dispute, chemical weapons disposal issues, and the Hanford site were not randomly selected, but had been concerned about the issues for a long time before collaborative analysis became an option. Whether this will present problems of legitimacy remains to be seen.

In addition, it may be difficult to conduct collaborative analyses on issues at a national scale. While Futrell's study gives some hope for this prospect, there were clearly defined local groups and national organizations with which to work. For more widely applicable public policies, there may be too many (or too diffused) interested parties for this to work. In these cases, consensus conferences may be needed to give a sense of the general public concerns that need to inform analyses. However, guidance on specific issues or problems that may arise in the process of research and analysis will be lacking.

I have also not addressed here the intricacies of process. Yet all the work that has been done thus far makes clear the importance of the details for how a process goes forward. How do participants get involved? How is their role delineated? Is there a mediator and how does s/he function? Who gets to set the agenda? The list of process issues is nearly endless and the details are often context dependent. Yet which process one uses should be shaped at least in part by the goals one embraces. By having a clear goal of maximizing public-expert interaction, ways in which to constructively support that interaction will hopefully become more apparent.

Finally, one might wonder whether there is a real possibility scientists will want to invite citizens into collaboration for policy-relevant analyses. As noted above in the discussion of science shops, part of the authority of science arises from the use of traditional scientific methods. Scientists must maintain a disciplinary integrity if they are to maintain some of science's authority in the policy realm. How are they to do this if the public is helping to guide their analyses? The answer lies in part in the philosophical discussion above: values should not replace evidence but rather should help make decisions under uncertainty. This, in practice, is a fine distinction, but it can be upheld through the use of a traditional scientific practice, peer review. If collaborative analyses are peer reviewed and found to uphold the expected high standards for methodological soundness, the authority of science, even in collaboration with the public, should not be undermined.

The need for contexts in which citizens can constructively debate scientifically-informed policy-making has never been greater. We need forums in which values relevant to these decisions can become clarified. As Futrell wrote: "It is through the expression of multiple concerns that we come to understand the common good of a diverse community that is central to good social decisions" (Futrell 2003: 475). Yet an expression of values that is not informed by the scientific or technical details is often just as irrelevant to decision-making as an expert's expression of personal values. Good community discourse is helpful, but it is even more helpful when it is soundly informed. Collaborative analysis has the added benefit that involved citizens gain an increased appreciation for the intricacies of scientific study and the analyses relevant to their communities. It is with these hopes that scientific experts may be persuaded to open the doors to the realm of science for increased public scrutiny *and* collaboration.

University of Tennessee, USA

NOTES

[1] By 'public' here I follow Dewey's notion that a public constitutes itself only when citizens recognize that they have an interest in something and thus come to form a public. When citizens recognize that there will be consequences that will affect them, a public constituency is formed. (Dewey 1927, chap. 1)

[2] Another critique of that standard view, that epistemic values cannot be clearly delineated from non-epistemic values, has been made by Rooney (1992) and Longino (1996).

[3] If methodologies don't improve, the judgments needed to do science often become 'standard practice,' thus erasing the appearance of the need for value considerations. Yet the values that shaped the initial judgments remain an influence through the practices accepted as standard.

[4] The NRC does not suggest that citizenry take up the charge of performing studies, data collection, or statistical analyses themselves, in contrast with the example of 'popular epidemiology' from Woburn Massachusetts (Fischer 2000: 151–7).

[5] As Renn describes it: "The objective is to reconcile conflicts about factual evidence and reach an expert consensus via direct confrontation among a heterogeneous sample of experts" (Renn 1999: 3050).

REFERENCES

Boiko, P., R. Morrill, J. Flynn, E. Faustman, G. van Belle, and G. Omenn (1996), 'Who holds the stakes? A case study of stakeholder identification at two nuclear weapons production sites', *Risk Analysis* **16**, 2: 237–249.

Busenberg, G. (1999), 'Collaborative and adversarial analysis in environmental policy', *Policy Sciences* **32**: 1–11.

Dewey, J. (1927), *The Public and Its Problems*, Athens: Swallow Press.

Douglas, H. (2000), 'Inductive risk and values in science', *Philosophy of Science* **67**: 559–79.

Douglas, H. (2002), 'The moral responsibilities of scientists: Tensions between autonomy and responsibility', *American Philosophical Quarterly* **40**, 1: 59–68.

Douglas, H. (2004a), 'The irreducible complexity of objectivity', *Synthese* **138**, 3, 453–73.

Douglas, H. (2004b), 'Border skirmishes between science and policy: Autonomy, responsibility, and values', in G. Machamer and P. Wolters (eds.), *Science, Values, and Objectivity*, Pittsburgh: University of Pittsburgh Press, forthcoming.

Fiorino, D. (1990), 'Citizen participation and environmental risk: A survey of institutional mechanisms', *Science, Technology, and Human Values* **15**, 2: 226–43.

Fischer, F. (1993), 'Citizen participation and the democratization of policy expertise: From theoretical inquiry to practical cases', *Policy Sciences* **26**: 165–87.

Fischer, F. (2000), *Citizens, Experts, and the Environment: The Politics of Local Knowledge*, Durham: Duke University Press.

Futrell, R. (2003), 'Technical adversarialism and participatory collaboration in the U.S. chemical weapons disposal program', *Science, Technology, and Human Values* **28**, 4: 451–82.

Gerber, M. (1997), *On the Home Front: The Cold War Legacy of the Hanford Nuclear Site*, Lincoln: University of Nebraska Press.

Guston, D. (1999), 'Evaluating the First US Consensus Conference: The impact of the citizen's panel on telecommunications and the future of democracy', *Science, Technology, and Human Values* **24**, 4: 451–82.

Irwin, A. (1995), *Citizen Science: A Study of People, Expertise, and Sustainable Development*, London: Routledge.

Joss, S. (1998), 'Danish consensus conferences as a model of participatory technology assessment: An impact study of consensus conferences on Danish parliament and Danish public debate', *Science and Public Policy* **25**, 1: 2–22.

Joss, S. and J. Durant (eds), (1995), *Public Participation in Science: The Role of Consensus Conferences in Europe*, London: Science Museum.

Kaplan, L. (2000), 'Public participation in nuclear facility decisions', in D.L. Kleinman (ed.), *Science, Technology, and Democracy*, Albany: SUNY Press, pp. 67–83.

Kinney, A. and T. Leschine (2002), 'A procedural evaluation of an analytic-deliberative process: The Columbia River comprehensive impact assessment', *Risk Analysis* **22**, 1: 83–100.

Kuhn, T. (1977), 'Objectivity, value, and theory choice', in T. Kuhn (ed.), *The Essential Tension*, Chicago: University of Chicago Press, pp. 320–39.

Lacey, H. (1999), *Is Science Value Free? Values and Scientific Understanding*, New York: Routledge.

Laird, F. (1993), 'Participatory analysis, democracy, and technological decision-making', *Science, Technology, and Human Values* **18**: 341–61.

Longino, H. (1996), 'Cognitive and non-cognitive values in science: Rethinking the dichotomy', in L. Hankinson-Nelson and J. Nelson (eds.), *Feminism, Science, and the Philosophy of Science*, Dordrecht: Kluwer, pp. 39–58.

National Research Council (1983), *Risk Assessment in the Federal Government: Managing the Process*, Washington, D.C.: National Academy Press.

O'Connor, J. (1993), 'The promise of environmental democracy', in R. Hofrichter (ed.), *Toxic Struggles*, Philadelphia: New Society Publishers, pp. 47–57.

Ozawa, C. (1991), *Recasting Science: Consensual Procedures in Public Policy Making*, Boulder: Westview Press.

Renn, O. (1999), 'Model for an analytic-deliberative process in risk management', *Environmental Science and Technology* **33**, 18: 3049–55.

Renn, O., T. Webler and P. Wiedemann (eds), (1994), *Fairness and Competence in Citizen Participation: Evaluating Models for Environmental Discourse*. Boston: Kluwer.

Rooney, P. (1992), 'On values in science: Is the epistemic/non-epistemic distinction useful?', in D. Hull, M. Forbes and K. Okruhlik (eds.), *Proceedings of the 1992 Biennial Meeting of the Philosophy of Science Association*, **2**, East Lansing: Philosophy of Science Association, pp. 3–22.

Rosenstock, L. and L. Lee (2002), 'Attacks on science: The risks to evidence-based policy', *American Journal of Public Health* **92**, 1: 14–18.

Rowe, G. and L. Frewer (2000), 'Public participation methods: A framework for evaluation', *Science, Technology, and Human Values* **25**, 1: 3–29.

Sclove, R.E. (1995), *Democracy and Technology*, New York: Guilford.

Sclove, R.E. (2000), 'Town meetings on technology: Consensus conferences as democratic participation', in D. Kleinman (ed.), *Science, Technology, and Democracy*, Albany: SUNY Press, pp. 33–48.

Stern, P.C. and H. Fineberg (eds), (1996), *Understanding Risk: Informing Decisions in a Democratic Society*, Washington DC: National Research Council, National Academy Press.

Wachelder, J. (2003), 'Democratizing science: Various routes and visions of Dutch science shops', *Science, Technology, and Human Values* **28**, 2: 244–73.

CHAPTER 10

Simon Joss

Between Policy and Politics

*Or: Whatever Do Weapons of Mass Destruction Have to Do With GM
Crops? The UK's GM Nation Public Debate as an Example of
Participatory Governance*

Introduction

The recent transformation in democracy, characterised by the emergence of new transnational systems of political and economic governance, according to Robert Dahl poses a fundamental 'democratic dilemma' between increased system effectiveness and citizen participation (Dahl 1994). On the one hand, the capacity for effective decision-making at large scale can be significantly increased through transnational governance systems, such as the European Union, the World Trade Organisation and the United Nations. On the other, this comes at the cost of direct influence of citizens on the processes of decision-making.

However, large-scale systems of governance transcending the control of the nation state and its citizens are arguably only one dimension of the third[1] historical transformation in democracy and its accompanying 'democratic deficit.' Another dimension is the widely perceived increasing complexity of issues having to be dealt with in governance processes involving a multitude of policy-makers, experts and stakeholders, and the related context of uncertainty within which decisions have to be made in the public interest (see, for example, Fisher 1999; Taylor 2004). Recent examples of the latter dimension include the issue of global climate change and technological innovations in agriculture, such as GM foods, and biomedicine, such as human cloning.

Thus, the democratic and social *problematique* of contemporary multi-level governance is concurrently characterised by the vertical dimension of (spatial) scale – involving different, often overlapping levels of decision-making, from the local, national, regional to the global – and the horizontal dimension of (thematic) complexity – involving contested expert knowledge, different socio-cultural practices and competing normative preferences. Furthermore, multi-level governance is increasingly characterised by new relationships between public and private actors, such as public-private partnerships (PPP), that challenge traditional forms of political responsibility and public accountability in the provision of public services.

171

*Sabine Maasen and Peter Weingart (eds.), Democratization of Expertise? Exploring Novel Forms of
Scientific Advice in Political Decision-Making – Sociology of the Sciences, vol. 24, 171–187.*
© Springer Science+Business Media B.V. 2009

The frequent references to the remoteness of contemporary decision-making, therefore, do not only relate to the physical distance between citizens and the political institutions representing them, but also to the communicative distance between the various expert discourses dominating technocratic policy- and decision-making and 'lay' discourses within the wider public sphere.

In response to this apparent democratic deficit and the related lack of legitimacy, there has been a growing body of scholarly literature to consider how contemporary public policy- and decision-making could be reconnected with citizens and the wider public through various forms of 'participatory governance' (see, for example, Kooiman 1993, Pierre and Peters 2000, and Grote and Gbikpi 2002). Dahl (1994) proposes the strengthening of democratic institutions and practices at national and sub-national levels, so as to improve democratic control over, and the delegation to, transnational decision-making. Others postulate the direct and regular involvement of social actors representing different types of expertise and special interests, as well as actors representing the general public interest, to increase the opportunities for mutual accommodation of interests, as well as to generate trust and accountability among those who participate (Schmitter 2002). Such 'heterarchical' networking among state and non-state actors, it is proposed, could help to come to grips with the complexity, diversity and dynamics of recent socio-technological developments and related structural changes (Kooiman 1993).

This scholarly debate has been matched by programmatic commitments by policy-makers to work towards greater accountability and public involvement, as illustrated for example by the European Commission's 2001 White Paper on European Governance (European Communities 2001). At practical level, new modes of participatory governance have been explored in relation to various public policy issues, such as urban planning, environmental sustainability and health care.

One area where for some time now there has been considerable experimentation with new forms of public and stakeholder participation in policy-making is in science and technology (see, for example, Joss and Bellucci 2002; Banthien et al. 2003). The reason for this lies in the often problematic relationship that has existed between politicians, experts and members of the public in relation to significant public controversies on science, technology and the environment, such as nuclear energy, information technologies, genetic modification and human reproductive medicine. New methods of 'participatory' and 'interactive' technology assessment (TA) and 'public engagement' – including so-called 'scenario workshops,' 'consensus conferences' and 'citizens panels' – have been implemented in various institutional and national settings, so as to render policy procedures socially more robust and politically more legitimate through more sophisticated socio-technological assessment and greater openness.

However, mirroring the contested nature of the issues considered within such participatory TA – which have ranged from transgenic animals, urban sustainability, information technology, radioactive waste management to gene therapy – the procedures themselves have often been subject to critical debate about their relative merit as tools for policy analysis and decision-making. Their role is often seen as ambiguous, owing to their dualistic function as assessment tools – a *quasi* 'extended expert peer review' process (Fixdahl 1997) – and as public policy-making fora – a *quasi* 'court of public opinion' within institutional settings. Criticism is variably raised on empirical-analytical ground, for example questioning the representativeness of participants, the framing of issues and the validity of outcomes; as well as on normative-conceptual ground, for example challenging their underlying political aims and strategies as well as democratic rationale.

This article analyses one such recent initiative of participatory TA, the *GM Nation?* public debate that took place throughout summer 2003 on the initiative of the Agriculture and Environment Biotechnology Commission (AEBC), an advisory body, whose remit is to advise the UK government on GM crops and food policy. The *GM Nation?* initiative lends itself for analysis, as it represents an interesting methodological extension of participatory TA in that it combined 'top-down' elements of public participation within a formal setting of policy-making with wider 'bottom-up,' informal processes of citizen involvement and public debate. Furthermore, as it was set in the wider context of the ongoing public controversy on GMOs that had erupted in Britain in the late 1990s, it allows for the analysis of the interrelationship of structured participatory procedures and wider socio-political processes. Thus, the *GM Nation?* is an ideal case study to critically assess, and reflect on, the practical manifestation of 'participatory governance' as a response to the perceived 'democratic dilemma.' The analysis is based on a combination of semi-structured interviews, participant observation and documentary analysis.[2]

THE *UK GM NATION?* INITIATIVE

Background: The 'Great GM Debate'

The *GM Nation?* initiative was ultimately the result of the 'great GM debate' that had swept across Britain in the late 1990s. Following a relatively quiet period in the early to mid 1990s, in which the controversy about GMOs and GM food had by and large been confined to the scientific and regulatory spheres with only occasional media coverage and limited public debates, from 1998 onwards the controversy magnified, spilling into the wider public sphere and rapidly becoming a major issue of political and public debate (see, for example, Gaskell et al. 2001; Weldon and Wynne 2001). There were a series of 'trigger events' that fuelled the controversy, against the backdrop of similar controversies having emerged in other European countries, and an already sensitive British public haunted by the BSE (bovine spongiform encephalopathy - 'mad cow disease') epidemic in the 1990s that had shaken British agriculture to its core and seriously undermined public trust in government policy and regulation.

Two such trigger events in the early phase were: firstly, the announcement in spring 1998 by *Iceland*, a major retailer, to ban GM ingredients from its own products (publicly referring to GM products as 'Frankenstein foods'), and to challenge US distributors to separate GM soybean from non-GM soybean; and secondly, the disclosure in summer 1998 in the *Observer* Sunday newspaper of controversial research findings by Dr. Arpad Pusztai at the leading public Rowett Research Institute in Scotland, which apparently indicated that GM potatoes fed to rats had shown adverse side-effects on the rats' intestines and immune system.

174   SIMON JOSS

There was widespread media coverage of these stories. The Rowett Institute's decision to terminate Dr Pusztai's contract and confiscate his research was portrayed by the media as an attempt to gag a reputable scientist and to prevent public scrutiny of the issue involved. The publication of statements both against and in favour of Dr Pusztai's research by different groups of scientists further fuelled the controversy. The GM debate entered the UK parliament in early 1999, where the leader of the opposition challenged the Prime Minister to introduce a moratorium on the commercialisation of GMOs. The Prime Minister retorted by complaining about the 'hysteria of public reaction,' the 'extraordinary campaign of distortion' by parts of the media, and 'the tyranny of pressure groups' (Moore 2001).

The controversy further intensified, with two tabloid newspapers (the *Daily Mail* and the *Express*) launching anti-GM campaigns. In summer, the Prince of Wales, a longstanding campaigner for organic agriculture, entered the fray with his opposition to GM food, publishing ten questions addressed to government and the wider public about the safety and usefulness of GMOs in the *Daily Mail* and setting up an Internet discussion group, which reportedly received tens of thousands of messages. By autumn 1999, various retailers withdrew GM products from their shelves. With the controversy showing little sign of abating, the government policy was diametrically pitched against public demands for a moratorium on GM crops by a broad coalition of media (from both left and right) and a growing network of civil society organisations, the latter forming the so-called Five-Year-Freeze' (FYF) network, which included diverse groups, such as the traditionally conservative, middle-class Women's Institute and Townswomen's Guild, and various environmental organisations as well as retailers. More radical direct action groups, such as Genetix Snowball went further by demanding an outright ban on GM crops. Successive opinion surveys showed significant public opposition to GMOs.

Finally, in early 2000, the government signalled a U-turn in its policy on GM crops (some commentators calling it the biggest U-turn since the Blair government had come to power in 1997). The Prime Minister for the first time publicly conceded, in the *Independent on Sunday* (27 February 2000), that there was 'cause for legitimate public concern' which the government understood well, and stated that 'consumers and environmental groups [had] an important role to play' in finding answers to the questions raised about GMOs. He explained that the government had 'radically overhauled the regulatory and advisory processes so that consumers have a real say on GM foods' and that confidence in the regulatory system would be restored by making it 'open, transparent and inclusive.'

Regulatory Streamlining and Opening-Up

In spring 1999, the UK Parliament through its House of Commons Select Committee on Environmental Audit recommended a new 'strategic' policy approach to GMOs, following a consultation that had shown the biotechnology regulatory framework to be too fragmented, lacking transparency, having too narrow a remit and not sufficiently representing civil society interests (ENDS 1999). This recommendation was followed through by the government with the announcement in summer 1999 to set up two new strategic commissions to advise on policy alongside the new independent

Food Standards Agency (FSA) – namely, the Human Genetics Commission (HGC) and the Agriculture and Environment Biotechnology Commission (AEBC). In addition, the government decided to set up a comprehensive system of field trials, so-called 'farm-scale evaluations' (FSEs), to compare herbicide-tolerant GM crops (corn, sugar beet and oilseed rape) with equivalent non-GM crops in terms of the effects of weed management on selected insect species (such as butterflies), the results of which were to be published in autumn 2003.

However, it should be noted that these three new commissions did not fully replace the various other already existing advisory bodies with more narrow remits and statutory powers, such as ACRE, the Advisory Committee on Releases into the Environment (Hails and Kinerlerer 2003). Rather, they were given an overarching position to provide strategic advice on all aspects of biotechnology relating to agriculture, the environment and human health. While the FSA is a statutory body with executive decision-making powers, both the AEBC and HGC are non-statutory bodies with non-binding advisory functions, a fact that was criticised in a report by the House of Commons Select Committee on Science and Technology (House of Commons, 12 March 2001).

The membership of AEBC – which became operative in summer 2000 with a budget of around GBP 100K – was opened to GM-critical civil society actors and experts[3]. AEBC's remit included: the consideration of wider social and ethical aspects of gene technology; the regular involvement and consultation of stakeholders and the public; and operation in accordance with criteria of openness, transparency, accessibility and exchange of information (URL: http://www.aebc.gov.uk). As a result, minutes of most meetings together with working documents and reports are published on the commissions' websites, meetings themselves are advertised and held in public in different parts of the country; and AEBC members are available for information to the public.

Public Involvement in GM Policy

Within just over a year of taking up its work, the AEBC recommended in its report *Crops on Trial* (AEBC 2001) public involvement in the decision-making process on the commercialisation of GM crops, stating that the government had approved the FSE field trials without providing the public with adequate information. In order to assess the (public) uncertainty surrounding GM crops and render the policy-making process more accountable, the report called for

> ... the facilitation of a broader public debate ... to foster informed public discussion. ... Whatever decisions are ultimately reached, they will be more palatable if they have not been taken behind closed doors. At present, there are no avenues for a genuine, open, influential debate with inclusive procedures, which does not marginalise the reasonable scepticism and wide body of intelligent opinion outside specialist circles. We need to harness new deliberative mechanisms ... in the form of a series of workshops, public debates and consensus conferences around the country (AEBC 2001).

In response, the government confirmed its commitment "to take public opinion into account as far as possible through an open decision-making process" (DEFRA 2002a), and asked the AEBC to elaborate a concrete proposal. The AEBC thus submitted a proposal for public involvement in GM policy to government in April 2002, following consultation with various stakeholders and specialists in public participation. The proposal recommended that the government should clearly set out the national and international legal context in which it would make decisions on GM crops and how it would take account of public views in making these decisions (AEBC 2002). Furthermore, it recommended that the involved public should be able to frame the specific issues (rather than the government), that enough time should be allowed to carry out the participatory initiative, and that the results of the FSE field trials should be fed into the public debate process.

The government's positive response stated that "the Government wants a genuinely open and balanced discussion on GM. There is clearly a wide range of views on this issue and we want to ensure all voices are heard" (DEFRA 2002b). The House of Commons Select Committee on Environment, Food and Rural Affairs, in scrutinising the AEBC's proposal and the government's response, emphasised the importance of maintaining the initiative's independence from governmental influence, and therefore endorsed the AEBC's proposal for an independent Steering Board to oversee the impartial implementation of the initiative.

In summer 2002, the government gave the official go-ahead for the *GM Nation?* initiative, for which the Prime Minister approved a budget of GBP 250K (falling well short of the requested GBP 1 million), setting the following conditions (DEFRA 2002b): in addition to the *GM Nation?* public debate, a parallel economic study (to consider the costs and benefits of GM crops) and a scientific review (to review scientific issues) were to be carried out, the former by the Prime Minister's Strategy Unit, the latter by a committee chaired by the Government's Chief Scientific Adviser and including the Department for the Environment, Food and Rural Affairs (DEFRA) and the Food Standards Agency (FSA). Furthermore, the *GM Nation?* initiative had to be completed before the scheduled completion of the FSE field trials in autumn 2003. Finally, the government made it clear that the governmental Central Office for Information (COI) should be in charge of implementing the initiative, which raised concerns in some quarters due to COI's close relationship with government and its relative inexperience with participatory procedures of this kind.

The government appointed Professor Malcolm Grant, the AEBC's chair, as chairperson of the Steering Board, which included six AEBC members (Bradley, Carmichael, Dale, Grove-White, Hann, Maxwell –see endnote 3) as well as three non-AEBC members – namely: Clare Devereux, director of Five Year Freeze (FYF – see above); Gary Kass, Parliamentary Office of Science and Technology (POST); and Stephen Smith, chair of the UK Biotechnology Council (an industry association).

In autumn 2003, the Steering Board convened a meeting of social scientists with expertise in science and society issues and experience of participatory governance to discuss the initiative. The invited group of social scientists subsequently criticised the narrow time schedule of the *GM Nation?* public debate, the lack of coordination between the three assessment strands, as well as the inadequate budget (Burgess et al. 2002). On behalf of the Steering Board, Professor Grant went public (both on radio and in the print media) with his criticism of the inadequate timeframe and financial resources (BBC 5 February 2003; Daily Mail 17 February 2003). This prompted the government to increase the budget by a further GBP 250K and to extend the time schedule until July 2003 (which was still before the publication of the FSE field trial results).

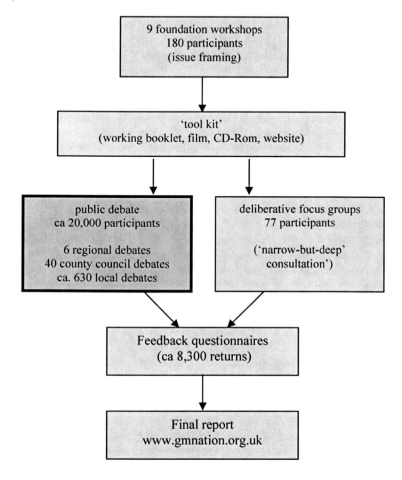

Figure 1: Methodological elements of the GM Nation? *initiative*

The *GM Nation?* initiative comprised several related methodological elements, as Figure 1 illustrates. The first methodological element was a series of nine so-called 'foundation discussion workshops,' each comprising 20 people, held in autumn 2002. For eight of these workshops, the aim was to allow members of the general public with no special or vested interest in GM crop technology to scope and frame the issues for the subsequent public debate. The participants were chosen from a random sample of members of the public, representing different age and socio-economic groups. The workshops, which took place locally across the UK, each lasted around three hours and resulted in the participants identifying the following six broad areas of interest: food, choice, (lack of) information, (lack of) trust, regulation, and commercialisation of GM crops, and the ethics of genetic modification. The ninth workshop was different in that it consisted of pro- and anti-GM campaigners. Together, these workshops resulted in the formulation of 13 questions, which formed the basis of the standard questionnaire used subsequently for participant feedback.[4]

The second element was a 'tool kit' outlining the issues at stake, based on the findings of the foundation discussion workshops. This consisted of various 'stimulus materials,' including: a 40-page working booklet (20,000 copies) to be used in the public deliberation; a CD-Rom (6,000); a film (1,100) distributed on video to broadcasters and shown in the public debate; and the *GM Nation?* website (URL: http://gmnation.org.uk). However, these materials only became available to the public shortly before, or during, the public deliberation phase, thus hampering timely public access and information provision. Even basic information, such as the dates and venues of the various debates, were hard to obtain from the organisers. Thus, the media (including newspapers, radio, television, as well as websites of various interest organisations) ended up being the main disseminator of information.

The main focus of the *GM Nation?* initiative was a series of public deliberation events taking place between 3 June and 18 July 2003. The Steering Board had initially allocated more time for this phase, but later had to limit it to a period of six weeks due to a combination of internal and external delays (decision to run foundation discussion workshops, request for additional funding, elections for the Scottish Parliament and Welsh Assembly). The shortage of time for the various public deliberation events was criticised by many participants, especially by those wishing to organise bottom-up events, as well as in view of the fact that the FSE field trial results were only going to be published later in autumn 2003.

The public debate comprised three tiers of deliberation: 'tier 1' consisted of six pre-structured, facilitated public meetings of 3-hours that took place at regional level in England (Birmingham, Taunton, Harrogate), Northern Ireland (Belfast), Scotland (Glasgow) and Wales (Swansea). Following the viewing of the commissioned film, participants in each meeting broke up into small-group sessions (comprising around 8-12 people), lasting around one hour, to discuss the issues and questions raised in the working booklet. In the second part of the meetings, each small group was asked to report their views and conclusions back to the plenum for further, facilitated discussion. Altogether, there were over 1,000 participants in these tier 1 meetings. Independent observers were invited to follow the proceedings and provide written feedback on the methodology, organisation and proceedings to the Steering Board.

'Tier 2' consisted of debates hosted by county councils (district authorities) in collaboration with the Steering Board. There were an estimated 40 meetings at this level. It was at times difficult to obtain timely information about where and when public meetings at this level were going to be hosted. (In several instances, the author received information from NGOs, because COI had no information to share.)

By far the largest number of events took place at 'tier 3' level, which comprised 'bottom-up,' locally organised meetings. According to COI's conservative estimate, some 630 events took place at this level (there may well have been more, as COI only counted those meetings where the local organisers had requested 30 or more feedback questionnaires and/or working booklets). These events were hosted by an array of local councils, research organisations, churches, environmental groups, galleries, villagers, and many *ad hoc* groups.

All participants in tier 1-3 events were asked to complete the official questionnaire (see endnote 4) after the meetings, to provide feedback of participants' assessment and views. COI received over 8,300 questionnaires from these meetings. However, only just over 1 in 3 participants in tier 1 events returned their questionnaire. This, together with the fact that most 630 local events can be assumed to have involved at least 30 participants, leads to a conservative estimate of around 20,000 participants in total. Others have estimated the number of participants nearer 35,000. The *GM Nation?* website registered over 14,000 questionnaire returns online, although these may include repeat completions. Finally, COI also received over 1,200 letters and emails from participants.

The feedback questionnaires were analysed by COI for inclusion in the Steering Board's final report (2003). They showed that only two percent of respondents found GM crops acceptable in any circumstances, whereas the vast majority of people cautioned against any hasty commercialisation of GM crops before sufficient risk and ethical analysis was carried out, and demanded proper safeguards.

In order to further verify the results of the various public debates – it was assumed that participants in the public debate were mostly people with particular interests in GM crops, rather than representing 'average' members of the public – the Steering Board commissioned a parallel 'narrow-but-deep' consultation, in the form of a series of deliberative focus groups involving 77 members of the public. These were carried out by the same private consultants that had organised the foundation discussion workshops. The focus groups consisted of two meetings: the first served to introduce the issues at stake and discuss the working booklet; the second, held two weeks afterwards, served to discuss participants views and concerns. The feedback questionnaires were completed at the beginning and the end of this process, so as to provide a before and after snapshot of participants' views. The results showed that, with more information available, the participants had become more sceptical and expressed greater concern about the various risks assessed. On balance, they shifted towards an anti-GM view, favouring a more cautious approach. Overall all, however, they were less pronounced in their opposition to GM crops than the participants in the public debates, and saw some benefits of GM crops (cheaper food, medical benefits, advantage for developing countries).

The final methodological element was the Steering Board's detailed final report, including a description of the methodology, summaries of the various events, and an analysis of the findings (Steering Board 2003). It was published in autumn 2003 and submitted to the government for consideration. The government published its written response in March 2004, signalling that within the government's overall strategy (announced in February 2004) for recommending the commercialisation of GM crops on a regulated, case-by-case basis, it would push for proper labelling of GM products, introduce measures to prevent cross-contamination of non-GM crops ('coexistence'), consider setting up GM-free agricultural zones, and providing information openly and transparently (DEFRA, 9 March 2004).

BETWEEN POLICY AND POLITICS

When considering participatory governance as a means of addressing the perceived remoteness and legitimacy deficit of formal policy-making institutions and processes, several dimensions are relevant for analysis, including: the substantive dimension, concerning the effect of a given participatory governance procedure on policy contents and outcomes; the instrumental dimension, regarding the nature of utilisation of the participatory procedure; and the normative dimension, concerning the meaning and values attached to the procedure. Importantly, these dimensions should not just be considered in respect of the main, dominant actor behind the participatory procedure, but also in respect of various other political and social actors that relate to the procedure in some way (e.g., as participants, observers, opponents). The particular characteristics of these dimensions and their interrelationship within a given participatory governance procedure make up its political and social relevance in terms of policy-making, public discourse and social experience.

Such an analysis leads to a number of critical observations in the case of the *GM Nation?* initiative. While this initiative arguably represents a bold and groundbreaking innovation in GM policy-making in Britain, it at the same time points to several weaknesses as a practical manifestation and model of participatory governance.

These weaknesses became apparent, and indeed a point of ongoing discussion, in the course of the various public events (Tier 1-3), as well as the related wider media and public debate. The deliberation at the regional event in Glasgow (Tier 1), as well as at of the local (Tier 3) events in Forest Row, a rural village in East Sussex, on 27 June 2003, were a case in point: there was repeated criticism of the timing, organisation and funding of the initiative. Why had only so little time been allocated for the public debate, several participants asked, given the importance of the issue? The six weeks available made it difficult for members of the public and interest groups to get their own local events up and running in time, especially as the provision of information in the planning phase proved inadequate. For example, getting through to the COI often proved difficult in the absence of comprehensive and timely information on the dedicated website. Another criticism was that little assistance was provided for organisers of Tier 3 events. In particular, there was no financial support available, not even for basic expenses, such as reimbursing the travel costs of invited expert speakers. In other words, people wishing to get involved in the debate, either as participants or organisers of their own events, often found this practically difficult. The high number of Tier 3 events, therefore, can be seen as a particular achievement by their organisers and is an indication of the social mobilisation potential at the time concerning GM crops among significant sections of the public across Britain.

Concerning the contents of deliberation, there was criticism of the framing of the Tier 1 and 2 events, which were based on the video shown at the beginning of the deliberation, worksheets summarising possible risks and benefits of GM crops, and the working booklet. Participants criticised the contents of the latter for presenting some arguments (as opposite 'pros' and 'cons') in what was thought to be rather simplistic ways. Also, they queried the compilation process. The booklet did not attribute sources, and did not explain that the contents were based on the foundation discussion workshops involving members of the public. Furthermore, as they were not available until the start of the debates, participants were not able to study them in depth. Thus, they did not fully serve as basis for the deliberation, as intended. There was further criticism of the fact that the results from the FSE (published in October 2003) as well as the economic and scientific assessments (published mid July 2003) were not availabe for consideration in the public debate.

What also became apparent during the discussions was that participants wanted to discuss the wider politics of GM crops in addition to the various more specific policy issues (such as risk assessment, regulation, labelling).[5] For example, why was there such an apparent rush to go ahead with the commercialisation of GM crops, given the many uncertainties involved, a participant at the Forest Row event asked? Others wondered what was driving the political process behind GM technology. Was it already a forgone conclusion? Was the UK government being pushed into promoting GM technology by the USA? Was Europe's precautionary approach with its *de facto* moratorium on commercialisation threatened by the overmighty, unaccountable World Trade Organisation and multinational companies?

Another revealing example was the comparison made at the Forest Row event between GM crops and the issue of weapons of mass destruction (WMD) in connection with the war in Iraq, which was a highly sensitive issue in the public domain at the time. Participants complained that, like WMD, it was difficult to know whether the government made all information publicly available, and whether scientific information had been manipulated to suit political decisions. They also wondered aloud whether government was paying proper attention to public opinion. The chair of the debate at first tried to steer the debate back to the issue of GM crops, insisting that the discussion was not about the war in Iraq and WMD. However, in the course of debate, participants returned to the comparison made. For example, one farmer said, to loud applause:

> My main concern tonight is that – this point of Iraq, actually – that whatever we are discussing, it's going to be of no consequence as far as the decision that's going to be made about growing GM crops [is concerned]. And that to me creates great anger, just like it did over Iraq, that the population can have one view and regardless of that the government goes ahead and does something else. And I see this as exactly the same as GM as well (Forest Row GM debate, 27 June 2003).

The group of local citizens hosting the Forest Row debate subsequently organised two 'bare witness'[6] events, during which participants stripped naked in a field of GM crops in East Sussex, and later in Parliament Square in London, to draw (media) attention to the issue of GM crops. Both events were reported in news bulletins on radio and television, and pictures printed in newspapers.

The point about how the results of the *GM Nation?* debate were going to be used was also forcefully made toward the end of the Glasgow event, when a participant, again to loud applause, said that while the deliberation had been useful, it remained doubtful whether the government was going to take the findings seriously.

This points to an ambiguity of the *GM Nation?* initiative, as perceived by many participants and observers, in terms of how the initiative fit into the formal policy-making on GM crops, as well as what the government's real intention behind, and attitude toward, the *GM Nation?* initiative was. This can partly be explained by the novelty of the initiative itself, and partly by its organisational setting: the AEBC is a relatively new agency charged with giving overall strategic advice (taking into account social and ethical aspect of GM technology) without, however, having any statutory function in policy-making, in contrast for example to the Advisory Committee on Releases to the Environment (ACRE – which formally is in charge of giving binding advice on GM crop releases) and the Food Standards Agency (FSA). ACRE reportedly showed little interest in the *GM Nation?* initiative and had little contact with AEBC. The ambiguity was further reinforced by the government's rather vague commitment "to take public opinion into account as far as possible" (DEFRA 2002a).

Hence, there was considerable suspicion and reservation on the part of many participants and commentators on the role and significance of the *GM Nation?* initiative. For many, these were confirmed in early 2004 when the government announced the go-ahead for the commercialisation of GM crops, although the government emphasised that this was to be done on a cautionary, case-by-case basis and in an open and transparent manner. Nevertheless, GM-critics complained that the government had ignored the findings of its own *GM Nation?* initiative and wider public opinion. The government's case was not exactly helped by the leaking of internal government documents in the *Guardian* in early 2004, in which the Secretary of State and her officials at DEFRA discussed how to 'wear public opposition down' by 'solid, authoritative scientific argument' (Paul Brown, *The Guardian* 19 February 2004: 1).

Thus, the *GM Nation?* initiative exhibits a certain paradox: on the one hand, the initiative was embedded in, and carefully controlled by, formal policy-making. Important parameters, such as timing (the time available for hosting the public deliberations), funding (the resources available to support the various elements), and framing (the setting of the agenda, the writing-up of the findings, the parallel scientific and economic assessments commissioned) were set and controlled by government and the Steering Board, with manifest impacts on the course of public deliberation. One source close to the organisation complained (in June 2003) that the government tried to exert control over the implementation, requesting regular meetings between DEFRA and members of the (supposedly independent) Steering Board almost on a weekly basis. On the other hand, the status of the initiative in relation to policy-making, and the government's commitment toward the initiative, were non-binding and remained relatively unclear and vague throughout the process.

Another difficulty facing the initiative was the issue of public representativeness. For example, some media commentators criticised the public debates for being dominated by people who had already made up their mind – namely, mostly GM-opponents, but also pro-GM scientists – and thus did not truly represent public opinion. One journalist asked: "why on earth did the government not commission a large-scale opinion survey instead?" (David Curry, *Financial Times* 17 October 2003: 21). This arguably misses the point, as the aim of the initiative was, as part of the policy-making process, to assess public perceptions on GM crops on the basis of in-depth deliberation, and to consult members of the public pro-actively and openly, rather than carrying out a closed, anonymous opinion survey with a statistically representative sample of average (and by – questionable - implication, relatively uninformed) members of the public. It should, therefore, not be surprising if interested people wishing to engage on the issue of GM crop commercialisation – that is, farmers, scientists, environmentalists, consumerists etc – formed the majority of participants in the various events.

Also, it is arguably rather patronising to call onto civil society and the wider public to engage with controversial, complex socio-technological issues, such as GM crops, only then to react surprised when large numbers of people actually do show an interest, bringing to the debate informed viewpoints (of whatever shade). Furthermore, it would be misrepresenting the participants at the events (at least the ones attended by the author) to suggest they were all avid campaigners with set views and causes. A significant proportion of participants in both the Glasgow and Forest Row events showed an interest in the debate in their capacity as college students, farmers, mothers, villagers, pensioners and politically interested citizens. The Forest Row debate had been well advertised locally (with eye-catching placards placed along the roads leading into the village), it took place in the old hall in the centre of the village, it was attended by the local Member of Parliament, each a pro- and anti-GM expert, as well as representatives of the local media. There were well over 150 participants, with some people having to listen to the debate standing in the entrance because of the unexpectedly large turnout. The atmosphere was cordial and the debate good-natured. Thus, the event resembled much more closely the proverbial 'town hall meeting' than a one-sided campaigning event.

Nevertheless, the question of how to define, represent and canvass 'the public interest' in participatory governance procedures, such as the *GM Nation?* initiative, is an important one. The organisers did pay attention to this by clearly stating the nature of participation in the public debates, as well as by carrying out the additional 'narrow-but-deep' focus groups, so as to have comparative data for further validation the findings of the public debates. Furthermore, various newspapers carried out their own statistically representative surveys to compare the initiative's findings with general public opinion.

CONCLUSIONS

When, in 1994, Robert Dahl called for the strengthening of democratic institutions and practices at national level in order to tackle the 'democratic dilemma' arising from transnational decision-making, he did so without giving much detail about how this might be done conceptually and practically. Since then, a growing body of scholarly literature has begun to address the socio-political phenomenon of multi-level governance with its spatial dimension of local, national and global decision-making, and its thematic dimension of socially complex and contested (scientific-technological) policy-making under uncertainty. Participatory governance has been proposed as a possible way forward for tackling such multi-level issues, and for helping to ensure the legitimacy of decision-making institutions. The involvement of various stakeholders, and even the wider public, in policy deliberation and decision-making, is now widely – and sometimes rather uncritically – postulated. This has resulted in a plethora of practical innovations in stakeholder networking and public participation.

The *GM Nation?* initiative – with its roots in the public controversy about GM crops, its special position in formal policy-making, and its methodological characteristic of large-scale public participation – offers itself for analysis as a recent practical manifestation and model of such participatory governance.

Overall, the initiative represented a serious and bold attempt by its instigators (the AEBC) and organisers (the Steering Board) to respond to the growing public calls for a more open and participatory style of policy deliberation and decision-making on GM technology in the UK. This was done through a methodologically innovative and diverse approach to public engagement. However, the analysis of its substantive, normative as well as practical dimensions reveals several critical points. One was the limited time allocated to the public debate phase; another the inadequate provision of public information and lack of transparency, while yet another was the limited support available to the organisers of the local events. These largely arose because of the restrictive conditions imposed on the 'independent' Steering Board by the government in return for giving the go-ahead for the initiative. At the same time, the government only gave a weak and rather unspecific commitment regarding the use of the findings in the policy- and decision-making process. Thus, there was something of a paradox between the close governmental control exerted over the initiative, on the one hand, and the non-binding nature of the process and its outcomes in relation to government policy-making on the commercialisation of GM crops, on the other.

Furthermore, there was a certain disjuncture between policy and politics. As an initiative instigated from within the regulatory system, the official emphasis was largely on policy, with the participatory process aimed at informing policy-makers on public perceptions and opinion on the commercialisation of GM crops (with the government committed to 'listen' to the outcome). However, the emphasis of the deliberation within the various public events was not just on policy, but significantly also on the wider politics of GM technology and the government's stance on GM crops. This showed itself in the political nature of the discussion as well as the considerable social mobilisation, especially in connection with the large number of various local debates. People not only seemed to want to participate as providers of 'public opinion,' but also as politically and socially engaged actors in their own right, wishing to influence and co-determine the politics of GM crops.

In view of these points, some people must inevitably have felt disappointed by the *GM Nation?* initiative. There was considerable public criticism from several quarters. However, with the necessary distance, one can view these critical aspects as a reflection of the political and social reality of the initiative and its wider context. Different parties – from the government, the various participants, to the media – had different stakes in the initiative and the contested issue of GM crops, and thus brought their particular interest to bear on the initiative in their role as organisers, participants or commentators. Thus, the initiative was instrumentalised and politcised in different ways and for various purposes. In turn, this led to a critical meta-level discourse on the worthiness of the initiative during the deliberations.

This has more general implications for the conceptualisation and analysis of participatory governance. For one thing, one needs to pay close attention to the particular circumstances that give rise to participatory governance initiatives, and the contexts within which they are placed. For another, one needs to consider how various

BETWEEN POLICY AND POLITICS 185

actors relate to, and interact with, such processes. Finally, and importantly, one needs to sufficiently recognise the politics of participatory governance. Doing so should help to understand and consider its actual, and not just its normative, potential and limits, as a dynamic, diverse socio-political process, for addressing the 'democratic deficit' of multi-level governance.

University of Westminster, London

NOTES

[1] The first transformation in the history of democracy, according to Dahl (1994), can be traced back to the emergence of democracies in Athens and Rome of Antiquity, the second to the emergence of modern nation states from medieval city-states.

[2] This research was carried out as part of the European Commission-funded research project 'Public Accountability in European Contemporary Contexts'. The empirical data used is partly derived from the case study carried out by S. Joss, A. Mohr, and C. Parau (unpublished paper). However, the analysis is the author's own.

[3] The AEBC membership in 2003 included: Prof M. Grant, Provost and President University College London (chair); J. Hill, former Director of Green Alliance (deputy chair); A. Bradley, Consumer Affairs Director for the Financial Services Authority; H. Browning, organic farmer; Dr. D. Carmichael, farmer; Prof. P. Dale, Research Director, John Innes Centre Norwich; Dr. E. Dart, Chairman of Plant Bioscience Ltd; Dr. M. Freeman, Senior Researcher, Medical Research Council Laboratory of Molecular Biology; J. Gilland, President of Ulster Farmers Union/farmer; Prof. R. Grove-White, Director of the Centre for the Study of Environmental Change, Lancaster University; Dr. R. Hails, Principal Scientific Officer, Centre for Ecology and Hydrology Oxford; J. Hann, freelance broadcaster and writer; C. Iweajunwa, member of executive evaluation group for NHS Direct, and member of Partners Council for NICE; Dr. D. Langslow, former Chief Executive of English Nature; Prof. J. Maxwell, Former Director, Macaulay Land Use Research Institute; Dr. S. Mayer, Director GeneWatch UK; J. Thornton, environmental law barrister at Allen and Overy Solicitors; Dr. R. Turner, Chief Executive, British Society of Plant Breeders. Source: http://www.aebc.gov.uk/.

[4] The 13 closed questions (with answers ranging from 'strongly agree' to 'strongly disagree') are: (1) I believe GM crops could help provide cheaper food for consumers in the UK; (2) I am concerned about the potential negative impact of GM crops on the environment; (3) I believe that GM crops could improve the prospects of British farmers by helping them to compete with farmers around the world; (4) I am worried that this new technology is being driven more by profit, than by public interest; (5) I would be happy to eat GM food; (6) I think that some GM crops could benefit the environment by using less pesticides than traditional crops; (7) I think that some GM crops would mainly benefit the producers, and not ordinary people; (8) I don't think we know enough about the long-term effects of GM food on our health; (9) I believe that some GM non-food crops could have useful medical benefits; (10) I am confident that the development of GM crops is being carefully regulated; (11) I am worried that if GM crops are introduced it will be difficult to ensure that other crops are GM free; (12) I feel that GM interferes with nature in an unacceptable way; (13) I believe that GM crops could benefit people in developing countries (GM Nation? The Findings of the Public Debate reported by the Steering Board). NB. There was a further, open-ended question: (14) 'Under what circumstances, if any, would you find acceptable for GM crops to be grown in this country?' Additional space was given for further comments and views.

[5] Around the same time, in the House of Commons (the lower chamber of parliament), Joan Ruddock MP in her motion speech (17 July 2003) complained of the lack of political debate about GM crops: "I want to speak on genetic modification. I believe that, with the exception of Iraq, GM is the most im-

186 SIMON JOSS

portant issue the House faces. Tomorrow is the closing date for the Government's public debate on genetic modification. Six days ago, the strategy unit reported on the costs and benefits of GM crops...The science review is expected next week. However, we have not had a single debate on the Floor of the House about the momentous decision that could be taken in our name before the end of the year. The public remain hostile to GM, yet we, their elected representatives, are woefully unengaged in [the] debate..." (House of Commons, 11 November 2003).

6 The term 'bare witness' is a word-play, replacing the commonly used verb 'to bear witness' (to give testimony) with the adjective 'bare' (stripped off, naked).

REFERENCES

AEBC Agriculture and Environment Biotechnology Commission (2001). *Crops on Trial*, London: AEBC. See also URL: http://www.aebc.gov.uk

AEBC Agriculture and Environment Biotechnology Commission (2002), 'A debate about the issue of possible commercialisation of GM crops in the UK', *Report of the Public Attitudes Group*, London: AEBC. See also URL: http://www.aebc.gov.uk

Banthien, H., M. Jaspers and A. Renner (2003), *Governance of the European Research Area. The Role of Civil Society*, Bensheim: Institut für Organisationskommunikation. URL: http://www.ifok.de

Brown, P. (2004), 'GM crops to get go-ahead. Leaked papers reveal decision', *The Guardian* (19 February): 1.

British Broadcasting Cooperation (BBC 5 February 2003), 'GM debate to inlcude crop trials', http://news.bbc.co.uk/1/hi/uk_politics/2727969.stm.

Burgess, J., T. Anderson, D. Dallas, A. Irwin, S. Joss, C. Marris, J. Petts, S. Rayner, A. Sterling, T. Wakeford and B. Wynne (4 November 2002), *Some Observations on the 2002–2003 Public Dialogue on Possible Commercialisation of GM Crops in the UK*, London: University College London.

Curry, D. (2003), 'Squaring the transgenic crop circle', *The Financial Times* (17 October): 21.

Dahl, R. (1994), 'A democratic dilemma: System effectiveness versus citizen participation', *Political Science Quarterly* **109** (spring 1994): 23–34.

Daily Mail (17 February 2003), '"GM a danger for the future," says Meacher', Article by S. Poulter.

DEFRA (Department for Environment Food and Rural Affairs) (2002a), *UK Government, Scottish Executive and Northern Ireland Department of the Environment Response to Crops on Trial Report*, London: Department for Environment, Food and Rural Affairs.

DEFRA (2002b), *Public Dialogue on GM. UK Government Response to AEBC Advice Submitted in April 2002*, London: Department for Environment, Food and Rural Affairs.

DEFRA (9 March 2004), *The GM Dialogue: Government Response*, London: Department for Environment, Food and Rural Affairs.

ENDS (1999), *Government Still Struggling to Master the Biotechnology Agenda*, ENDS Report no. 292 (May 1999).

European Communities (2001), *European Governance. A White Paper*, Luxembourg: Office for Official Publications of the European Communities (ISBN 92-894-1061-2).

Fisher, F. (1999), 'Technological deliberation in a democratic society: The case for participatory inquiry', *Science and Public Police* **26** (5): 294–302.

Fixdahl, J. (1997), 'Consensus conferences as extended peer review', *Science and Public Policy* **24** (6): 366–76.

Forest Row GM Debate (27 June, 2003), For a full transcript of the debate, see URL: http://www.sussexgmforum.org.ukGM debate_transcript_june03.htm.

Gaskell, G., M.W. Bauer, N. Allum, N. Lindsey, J. Durant, and J. Leuginger (2001), 'United Kingdom: Spilling the beans on genes', in G. Gaskell and M.W. Bauer (eds.), *Biotechnology 1996–2000. The Years of Controversy*, London: Science Museum, pp. 292–306.

Grote, J.R. and B. Gbikpi (2002), *Participatory Governance. Political and Societal Implications*, Opladen: Leske & Budrich.

Hails, R. and J. Kinerlerer (2003), 'The GM public debate: Context and communication strategies', *Nature Reviews Genetics* **4** (October 2003): 819–25.

House of Commons (12 March 2001), *Fourth Report of the Select Committee on Science and Technology*, London: House of Commons.

House of Commons (11 November 2003), *GM Science Review*, Debate Pack, page 27 (referring to HC Deb 17 July 2003 Vol 409 c 504-6), London: House of Commons Library. URL: http://hcl1.hclibrary.parliament.uk/parliament/debatepacks.asp.

Independent on Sunday (27 February 2000), 'The key to GM is potential, both for harm and good', Article by Tony Blair.

Joss, S. and S. Bellucci (eds.), (2002), *Participatory Technology Assessment. European Perspectives*, London: Centre for the Study of Democracy (University of Westminster).

Joss, S., A. Mohr and C. Parau (2003), *The GM Food Controversy, the AEBC and the GM National Debate*, PubAcc Report, London: University of Westminster (unpublished paper).

Kooiman, J. (ed.), (1993), *Modern Governance. New Government – Society Interactions.* London, Thousand Oaks, New Delhi: SAGE Publications.

Moore, J.A. (2001), 'More than a food fight', *Issues in Science and Technology online.* URL: http://www.nap.edu/issues/17.4/p_moore.htm.

Pierre, J. and B.G. Peters (2000), *Governance, Politics and the State*, Basingstoke, London: Macmillan Press.

Schmitter, P. (2002), 'Participation in governance arrangements: Is there any reason to expect it will achieve "sustainable and innovative policies in a multilevel context?"', in J.R. Grote and B. Gbikpi (eds.), *Participatory Governance. Political and Societal Implications*, Opladen: Leske + Budrich, pp. 51–69.

Steering Board of the Public Debate on GM (Genetic Modification) (2003), *GM Nation? The Findings of the Public Debate. Report of the Steering Board of the Public Debate on GM (Genetic Modification) and GM Crops.* URL: http://www.gmnation.org.uk.

Taylor, I. (ed.), (2004), *Genetically Engineered Plants. Governance Under Uncertainty*, Binghamton, NY: Haworth Press.

URL: http://www.aebc.gov.uk.

URL: http://gmnation.org.uk.

Weldon, S. and B. Wynne (2001), *The UK National Report. Assessing Debate and Participative Technology Assessment (ADAPTA). Final Report*, Lancaster: Centre for the Study of Environmental Change, Lancaster University. See also ADAPTA final report at URL: http://www. inra.fr/Internet/Directons/SED/science-governance/pub/ADAPTA/.

CHAPTER 11

MATTHIJS HISSCHEMÖLLER

PARTICIPATION AS KNOWLEDGE PRODUCTION AND THE LIMITS OF DEMOCRACY

INTRODUCTION

This paper starts with a twofold observation: On the one hand, present day democracy and policy-making are confronted with a trend toward participatory policies rather than top-down policy-making. This trend is generally justified by the observation that current democracies face a crisis of legitimacy. There is widespread concern for a widening gap between the public and political agendas at national and international levels, including the European Union. Policy science has captured this trend in terms of a shift from 'government' to 'governance.' In 'governance,' power and responsibility have become dispersed among many actors at the national and international levels, but so is accountability. Traditional democratic institutions such as parliament are loosing power to institutions that serve to facilitate policy making by networks involving interested parties and expert communities. There is a growing awareness that issues such as the transition to a sustainable energy system, water management, food safety and the like, are characterized by the need for a long term approach, whereas elected officials have a rather short time horizon.

On the other hand, participatory tendencies in the area of political decision-making have become reflected in scientific practice that aims at producing knowledge for policy (Hisschemöller, Hoppe, Dunn and Ravetz 2001). The development of participatory knowledge production has been justified in several ways. Most importantly, today's complex issues cannot be effectively addressed from an academic point of view. Knowledge for policy would require interdisciplinary cooperation and 'extended peer review' in order to take account of the various goals and problem definitions of the stakeholders involved.

Although many have welcomed participatory developments in both policy and science as improvements of democratic practice, several authors have raised doubts. Critics assert that stakeholder participation weakens the policy-science boundaries and, in consequence, the integrity of both discipline-based science and democratic politics. Ezrahi (1990) has argued that participatory practice tends to undermine representative democracy in three ways: Firstly, by questioning the impartiality and objectivity of science, it has undermined the most powerful legitimization 'tool' of the liberal democratic system, the mechanism of depersonalizing the exercise of power through technical arrangements. Secondly, participation has brought about a

Sabine Maasen and Peter Weingart (eds.), Democratization of Expertise? Exploring Novel Forms of Scientific Advice in Political Decision-Making – Sociology of the Sciences, vol. 24, 189–208.
© Springer Science+Business Media B.V. 2009

shift in the focus of policy-making from a technical to a dramaturgical approach, replacing real interventions that serve political ends by symbolism and rhetoric. And thirdly, because of this, participation has undermined the transparency of policy-making and thereby the possibility to hold political decision-makers responsible and accountable for their actions. An other important objection to participation is that it may corrupt the integrity of science, given the many historic examples of misuse of science for ideological purposes. And even if liberal-democratic systems are supposed to possess an in built mechanism to prevent the worst cases from happening, the trend toward stakeholder involvement in research has raised legitimate questions with respect to possible misuses of science for sustainability and other political objectives. These criticisms deserve more serious attention than they have so far received in the (environmental) policy sciences community.

This paper sets out to, in a tentative way, unravel the complex relationship between participation, democracy and science. Although I endorse the claim that participatory trends in both politics and science are necessary and even inevitable, I will take argument with mainstream participatory discourse. The next section will unfold my central claim, which at the same time shapes the bias of the paper's argument: Whereas democracies have managed quite well to deal with conflicts of interests and of values, they have so far proven unable to effectively address conflicting knowledge claims at the level of the *political process* and the *political institutions*, i.e., the formal and informal rules of the game that shape political processes in democratic political systems. Then, I will analyze what I see as the key mechanisms which limit participation in policy processes in such a way that conflicting knowledge claims are organized out. Next, I will analyze mechanisms for reducing and enhancing conflict in participatory assessments meant to assist policy making. The concluding section will discuss in a tentative way how political institutions could be adjusted to effectively manage participatory knowledge production and how such adjustments may also help to resolve the apparent tension between participatory governance and representative democracy.

The paper's focus is on critical reflection, drawing on personal observations in combination with theoretical analysis. Rather than testing a given hypothesis through the collection of empirical data, the paper is meant to develop some explanatory hypotheses about the shortcomings of present day participatory discourse.

THE SOCIAL-CONSTRUCTIVIST CHALLENGE

The social-constructivist perspective stresses the notion that knowledge and the language used to conceptualize it, cannot be considered impartial or even objective, since the problems at stake are socially and politically constructed. In my view, this is not identical to the position that social reality cannot be known or, even more radical, does not exist. I take social constructivism as to acknowledge that social and political contradictions are a main feature of social reality itself, which not only affect peoples' values but also their 'facts.' The relevance of social constructivism is that it in a way extends the classic definition of a social problem as a gap between a given situation (the 'facts') and a desired one ('values'), because it points to the interplay of values and facts, stressing that different problem constructions cannot be simply reduced to 'value conflict.' The observation that 'facts' and knowledge claims do matter is the basic justification for the academic interest in lay or practical knowledge next to, or even in contrast to expert knowledge (e.g., Schön 1983).

The social constructivist position has been used to highlight the need for a participatory approach (e.g., Hajer 1995). However, this is not what in my view makes this position particularly challenging. The argument could also go the other way: If social problems imply constructions of reality, then everyone is entitled to his own problem construction and participation would not lead to 'better' policies. This argument can be found in the work of social constructivists avant-la-lettre, such as Schumpeter and Hayek, but has led them to quite different inferences. Schumpeter concludes that any form of participation in policy making should be avoided. Even the most innocent attempts to influence policies as writing letters to policy-makers may harm the integrity of statesmanship (1942). In contrast but for the same reason, Hayek (1944) has launched his frontal attack on the legitimacy of the state: He stresses that it is impossible to render any public cause from the infinite number of social constructions in a society. This leaves us with the observation that social constructivism may be used in defence of totally opposite positions with respect to the legitimacy of participation and political order. However, social-constructivist views, irrespective of their differences, have one observation in common: In contrast to political theories which, from an objectifying perspective, see political conflict especially as a conflict between values and interests, social constructivism stresses the importance of competing and conflicting *knowledge claims*.

This observation leads to the following hypothesis, which is the central claim of this paper: Democratic systems, which have evolved in the 20th century, may have proven quite capable of dealing with conflicts of *values* and *interests,* but they have proven unable to effectively manage *conflicting knowledge claims*. Conflicting knowledge claims, as the concept is used here, refers to scientific knowledge as well as lay or practical knowledge. I am far from saying that interests and values can be considered apart from knowledge. If social-constructivism is taken seriously, one must accept the assertion that *interests* and *values* articulate *knowledge* claims as much as *knowledge* articulates *values* and *interests*. Indifferent of how these concepts are defined, the fact that language allows them to co-exist implies that they refer to different things. Rather than a matter of definition, my point is that *political processes* have a preference for articulating conflicts of values and interests and suppressing conflicts related to competing knowledge claims. Whereas values and interests are treated as legitimate categories in political discourse, conflicting knowledge claims are often not taken for what they are, that is conflicting observations with respect to socio-political contradictions. And if so, their evaluation is not considered a matter of politics but as one of science.

Yet, the hypothesis cannot be taken as a denial of the critical importance of knowledge for the well-functioning of effective democratic systems. On the contrary, it fully concurs with Ezrahi's (1990) notion of the relevance of science and expertise for liberal democracy in that knowledge helps to instrumentalize and depersonalize the use of political power. This line of reasoning in a sense even supports my claim, as it is based on the widely shared assumption that political action requires consensus on the knowledge for policy.

THE BIASES OF PARTICIPATION

How do democratic systems manage to exclude conflicting knowledge claims from straightforward consideration?[1] Figure 1 below presents a meta-theory, which distinguishes different types of policy problems according to their *structure*, which is defined as the relationship between contents and process. It also indicates the impact of policy process on the role of science in public policy. Given the social-constructivist perspective, it should be noticed that the distinction between knowledge and values is ideal-typical. The figure's cells show that knowledge and values are always articulated in a specific way.

Consensus on relevant values? / Consensus on relevant knowledge?	NO	*YES*
NO	Unstructured Problem Policy as learning Science as problem finding A	Moderately Structured Problem Policy as negotiation Science as advocate B
YES	C Badly Structured Problem Policy as accommodation Science as mediator	D Structured Problem Policy as ruling Science as problem solver

Figure 1: Four types of policy problems and their bearing on the role of science in public policy

The typology conveys a twofold message to the policy analyst: First, it shows how actual policies may reflect a correspondence between contents and process. It is assumed that in case of correspondence the (conflicting) information for addressing the policy issue has gained access to the policy agenda. The second message of the typology is that it provides a clue of how to look for mechanisms of exclusion. Two general mechanisms are distinguished:

1. The policy process maybe based on the assumption that all relevant knowledge is taken into account, which assumption is wrong, and

2. The policy process maybe based on the assumption that the relevant values (e.g., problem frames, policy goals) are taken into consideration, which assumption is wrong.

The cells of the typology reflect arguments from theories on policy-making and democracy. These insights are now used to explore how conflicting knowledge claims may become excluded from the political process in specific cases of policy-making, thereby assuming that these mechanisms maybe more or less frequently observed in all democratic polities.

Policy as Ruling: The Privileged Position of Expert Knowledge

The structured problem leaves the actual decision-making to experts. Who is considered an expert is dependant on the issue's context. Experts can be physicists, doctors, lawyers, politicians or social workers. The decision-making agency has all characteristics of a classic bureaucracy. With respect to the affected stakeholders and the public at large, the decision-making agencies appear as a monolithic actor. Policy agendas do not allow for debating competing knowledge claims. Rival hypotheses maybe dealt with by the science community, they are not supposed to have a bearing on policy. This type of policy works as long as there is consensus on the technical character of the issues involved and the impartiality of the (scientific) experts.

Dahl (1985) and Fischer (1990) take argument with rule by experts, thereby referring to theories of guardianship, such as Saint-Simon's theory on good governance.[2] Such theories defend rule by virtue of certain qualifications. Both Dahl and Fischer consider the growing power of experts in certain policy areas a threat to pluralist democracy and a shift towards technocracy. However, it is equally defensible to take this style of policy-making as indispensable for a well-functioning liberal democracy, which uses science and expertise to instrumentalize and depersonalize, to use Ezrahi's (1990) expression, the exercise of political power.

How policy as ruling organizes out contradictory knowledge claims may be illustrated by the case of the UK. The UK political system is often cited as the best example of a majoritarian democracy where "the winner takes all" (Lijphart 1984). In this system, where the decision-makers hardly have the need to negotiate with an opposition or to form winning coalitions, a somewhat technical style of policy-making is likely to foster the legitimacy of the party in power. In her analysis of the British BSE scare, Jasanoff points to a system where status and integrity determine the attitude of the public with respect to the leading policy advisors, "a relationship founded on shared values and deference to expertise – which is increasingly at odds with the conditions of citizenship in the modern world" (Jasanoff 2001: 261). And: "In the British regulatory process, then, public confidence in governmental advisors is secured through testing the reliability of persons (rather) than (primarily) the rationality of their views" (idem). "Advisors often relay their conclusions to decision makers in confidence and reports, when they are published, are rarely backed by records of behind the scenes argument or dissent" (idem). It is not a custom to consult the public and everything is being done to prevent an adversarial process. Hence, in the (ideal typical) UK style the expert group, which constitutes and maintains itself, decides by virtue of its privileged position what 'knowledge' decisions maybe based on.

It could be argued that there is more to say to the UK mechanisms of dealing with participation and conflicting knowledge claims. Especially the *public inquiry* is considered an impressive participatory tool, where citizens arguments receive close attention (Huitema 2003). To some extent, indeed, the public inquiry makes it possible to seriously consider conflicting knowledge claims put forward by parties in an environmental controversy. So, even if the inquiry is considered imperfect given its bias for expert rather than lay knowledge, one may get the impression that through this instrument the UK political process has found a way to address competing knowledge claims after all. However, this is not entirely the case. Observers have pointed to the basic rule underlying the inquiry process, that the leading inspector has to decide what arguments are to be considered 'expert knowledge claims.' The inspector weighs the arguments put forward as if he were a judge. So, the public inquiry is to be considered a quasi legal process rather than an inherent part of the political process itself (Barker and Couper 1984; Huitema 2003). Hence, it is justified to conclude that the public inquiry system does not so much indicate that politics have found a way to deal with conflicting knowledge claims, but rather that politics have found a(n elegant) way of organizing rival hypotheses out of the political process.

As policy as ruling is built on closed and hierarchical networks of expertise, which have the privilege framing the information contents for policy makers, and given the built-in mechanism to avoid an open adversarial process, this type of policy is largely unable to address conflicting knowledge claims. Knowledge claims that contradict prevailing assumptions can gain access to the political agenda either by the election of new officials, replacement of staff and through changes in the expert networks themselves.

Policy as Negotiation: Shaping the Conflict of Interest

Policy as negotiation aims at finding a trade-off between conflicting interests. This policy type can be understood by reference to the typical American way of policy–making, as captured in concepts such "disjointed incrementalism" (Braybrooke and Lindblom 1963) and "partisan mutual adjustment" (Lindblom 1965). In contrast to Schumpeter's pluralist model, the US model considers stakeholder participation (lobbying) as a regular feature of the governmental process. It assumes that citizens organize in order to lower the costs of participation and maximize the opportunity to achieve their political goals. However, this pluralist conception also assumes – and this is critical for the stability of the system – that (1) the political elite reflects the heterogeneity of the electorate, which (2) to some extent guarantees alternate majorities, (3) that social interest groups overlap which contributes to the sharing of basic values and to (4) a melioration of positions. In short, social heterogeneity is vital for this model, as democracy then "makes for enough consensus to hold the system together and enough cleavage to make it move" (Berelson et al. 1954: 318).

How does this system work with respect to participation and the recognition of competing knowledge claims? Advocacy coalitions (Sabatier and Jenkins Smith 1993) use science to strengthen their position and weaken the claims of other coalitions. The 'knowledge in use,' 'practical knowledge' of 'cognitive maps,' terms used to refer to policy makers' assumptions, filters and selects the scientific information which is or can be made consistent with pre-existing views and insights. If this policy process works well, there is some possibility that at a certain point in time conflicting knowledge claims are explored and discussed within and across coalitions, i.e., *policy oriented learning*. But learning takes quite some time, according to Sabatier at least a ten year period. Furthermore, advocacy coalitions are only open to explore information that contradicts their own assumptions when they are under huge external pressure, e.g., disasters or loss of public support. Therefore, like in policy as ruling, 'learning' mostly occurs in an indirect way, through the election or appointment of new decision-makers and staff.

Negotiation is very different from ruling in handling expertise, as the US system is considered open, adversarial, formal and legalistic. In this open atmosphere, advisors are continuously subjected to supervision and challenge (Jasanoff 2001). US experts are in a less privileged position than their colleagues in the UK. Therefore, the US system does not have that many possibilities for ruling out competing knowledge claims by maintaining sharp policy-science boundaries. In contrast, in the process of Negotiation, it is the blurring of policy-science boundaries which leads to avoiding a reflection on competing realities and truth claims.

Knowledge claims are linked to (vested) interest positions. This happens, whether the experts involved like it or not. Knowledge claims that may articulate new or independent positions are either ignored or are translated into a warrant in support of an existing position. It may happen that scientific positions can be used to meliorate a conflict of interests, which may help to settle the dispute in an incremental manner. However, if information is ignored, the conflict may suddenly polarize. This can be illustrated by the many examples of so-called NIMBY behaviour, e.g., when local opposition against the siting of a facility associated with environmental or health risks is addressed as if it were based on clean calculation and self-interest. Decisions and decision-makers may have a problem of legitimacy, because the method used for arriving at decisions is only adequate insofar competing interests are at stake but not in case of conflicting knowledge claims. In order to overcome deadlock, at least one of the parties may seek for an institution that is able to take into account the *truth* as a value independent of (perceived) interest. Whereas the US system apparently lacks an ingenious instrument such as the UK public inquiry, a likely way is to go to court.

In conclusion, Negotiation has a particular capacity of handling political conflict, i.e., through shaping conflicting positions into interest positions. For this policy type to work, parties take a meliorative approach. This requires that preferences can be ranked on a single scale so that the acceptability of an option may increase through trade-off . However, when the conflict between knowledge claims tends to take over the conflict of interests, policy as Negotiation does not work anymore. Institutions that facilitate political decision-making, especially those that allow for an open pluralist process, may come under severe pressure. The more participation, the less likely is the possibility of a *political* settlement of the conflict. This is, because conflicting knowledge claims become intertwined with the articulation of advocacy positions and the process of majority formation instead of being taken for what they are, rival hypotheses with respect to the 'truth.'

Policy as Accommodation: Shaping the Conflict of Values

The Dutch political scientist Lijphart (1968) introduced the concept 'accommodation politics' in order to typify the Dutch political system as it had evolved during the first part of the 20th century (1917–1967). In later publications, this type of system has also been labelled the consensus model of democracy in contrast to the majoritarian model based on the 'winner takes all' principle. Accommodation politics differs strongly from American pluralism in that it applies to a social system, which is based on some sort of social segregation. In such context, a competitive pluralist approach would either yield oppression of minorities or the political system would fall apart. The democratic method fit for this particular situation would be a kind of elite rule based on a compromise among the leaders of the various cultural, ethnic or religious groups. This compromise would include critical rules of conduct, such as an agreement to disagree, which implies a *de facto* veto power for the blocks involved. Accommodation also involves secrecy vis-à-vis the rank and file. This model shares its rejection of participation with the model of expert ruling, as participation may destabilize of the political system. In one respect, however, Accommodation is very different from the pluralist models that are represented by both the UK and US, i.e., the absence of competition among the political elites (Huntington 1981).

Policy as accommodation may work in cases of irreconcilable values, such as culture, ethnicity or religion, it may also work in other controversies on environmental risk (Schwarz and Thompson 1990). After all, environmental conflict may articulate antagonistic values, like in the cases of nuclear power, GMOs or the protection of traditional landscapes and natural areas. The basic mechanism in Accommodation as a policy strategy is to seek consensus on means rather than ends. Means can be understood as all kinds of vehicles that may help to move away from a deadlock position, such as the conception of general policy framework documents that seek at integrating competing values (ecology versus economy, etc.) at a level so abstract that it does not (yet) touch the really hot potatoes, the application of broad policy principles, such as the precautionary principle, as well as concepts used to enhance dialogue and to establish a shared discourse, such as *sustainability, ecological footprint* or *transition management*. Policy may become symbolic in character. The basic idea is that a continuous dialogue among the parties may build trust and create a shared framework for understanding the complexities of the situation at hand.

It may not come as a surprise that the role of science and expertise is critical in Accommodation. The contribution by expertise tends to limit participation in two ways. Firstly, the level of abstractness of policy discourse and the scientific jargon discourage members of the attentive public to stay involved. Secondly, not only the number of participants is limited but also their role gets modified. Participants are expected to act as experts with respect to the perceptions, interests and values related to a certain issue. Through this subtle change in role the policy process will look distant as compared to Negotiation.

Policy as Accommodation is often used as an alternative for Negotiation, when an issue gets over-politicized. In order to understand what may happen with the science when the shift is made from Negotiation to Accommodation politics, this section first looks into the conclusions by Jasanoff in her famous study *The Fifth Branch* (1990). It is fair to state in advance that the concepts used by Jasanoff do not necessarily have the same meaning as they are given in this paper. However, her observations indicate that the US system of advisory boards successfully uses mechanisms that can be understood in terms of Accommodation. The first point to be made is that Jasanoff advises to avoid both the Scylla of technocratic science – policy separation and the Garybdis of politicization. In terms of policy types, what needs to avoided is both the models of expert ruling and negotiation:

> Scientific advice may not be a panacea for regulatory conflict or a failsafe procedure for generating what technocrats would view as good science. It is, however, part of a necessary process of political accommodation among science, society and the state and it serves an invaluable function in a regulatory system that is otherwise singularly deficient in procedures for informal bargaining. In order to accomplish this, science may need to negotiate some space to withdraw from politics where it can work out and negotiate 'serviceable truths.' In doing this, scientists get committed to moderate their views 'toward a societal mean' (Jasanoff 1990: 250).

What actually happens with competing knowledge claims in this process of accommodation and compromise? It is likely that antagonistic viewpoints are transformed into more abstract and general values. These values may play a visible role in political rhetoric but they are in fact organized out of the actual problem solving. The accommodation process may show that the science used in support of the advocacy positions is replaced by other types of expertise. An example is found in the study by Hoppe and Peterse (1993) on the controversy on LPG landing in the Netherlands (1980s) (see also Hisschemöller et al. 2001: 451–3). Accommodation has benefited the emergence of integrated methods such as risk analysis, impact assessment, technology assessment and integrated assessment. They are widely used to provide an interdisciplinary, basically quantitative (modelling) alternative for a process that is characterized by the articulation of rival scientific perspectives.

At the institutional level, science-policy interfaces have emerged that help in creating boundaries for legitimate policy science discourse, which happens especially by defining 'scientific uncertainty.' Policy science interfaces or epistemic communities have proven especially useful in facilitating political compromise in international environmental agreements, the International Governmental Panel on Climate Change (IPCC) being one of the most cited examples (e.g., Gupta 2001). The IPCC can neither be considered an open forum for debating conflicting positions in climate science nor as a closed expert community. Its major function is to shape common discourse with respect to incorporating the political sensitivities into the global climate change scientific reports.

In conclusion, policy as accommodation has this particular capacity of handling political conflict through transforming conflicting positions into values. Rather than debating values, parties focus on means that may provide a way out from deadlock. Consensus on knowledge is a prerequisite for political compromise. The risk of this policy type is that, by providing a pragmatic solution strategy for conflict between irreconcilable values, it creates institutions and discourses that, because of their strategic asset, get a vested interest in addressing value conflict. If conflicting knowledge claims are considered values rather than knowledge, accommodation may become an obstacle rather than a vehicle for problem solving, as critical hypotheses are not being explored and may even remain unnoticed. This observation raises questions with respect to the qualities of so-called integrated methods, including tools and procedures for participatory assessments. If applied in the context of an accommodation strategy, participatory exercises may have this unintended effect that they not only prevent conflicting knowledge claims from entering the political agenda but the scientific agenda as well.

Summarizing the Main Observations

This section has explored how types of democratic governance manage to organize out conflicting knowledge claims from political decision making. The main observations are summarized in Figure 2.

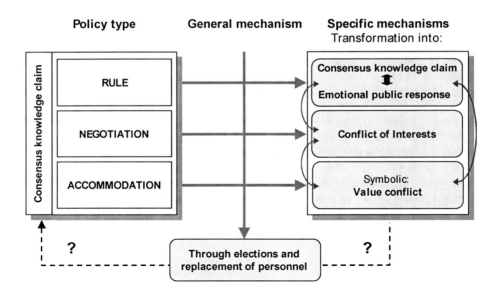

Figure 2: Mechanisms to deal with conflicting knowledge claims in three types of policy

The observations indicate that three types of policy that, according to a broad body of knowledge on politics and policy making dominate democratic political systems and policy-making institutions, tend to avoid conflicting knowledge claims from being openly considered as part of the political process. This is most obvious in policy as Rule, where self-established networks of competent experts define the boundaries of authoritative knowledge. Where participation is to some extent encouraged, i.e., in *negotiation* and *accommodation*, conflicting knowledge claims are transformed into conflicting interests or values, respectively. Hitherto marginalized knowledge claims may become dominant through personnel changes. Rather than the political process, (quasi) legal procedures may provide an opportunity to bring critical information to bear.

At this point, I would like to stress that these observations cannot be interpreted as an oversimplification of the political process. In democracies, conflicting knowledge claims are part of the day-to-day political debate. In some instances, they may even become subject of political inquiries that are explicitly aimed at evaluating the state of the art knowledge with respect to a specific issue. Policy learning by confronting rival claims happens. My point however is that this is an exception rather than a rule and that current democracies lack the institutions to facilitate participation as knowledge production rather than to express one's concerns, interests or values.

PARTICIPATORY KNOWLEDGE PRODUCTION: THE METHODOLOGICAL CHALLENGE

Whereas participation has for quite some time been associated with the realm of policy, the articulation and confrontation of competing knowledge claims is generally considered a task for science. Approaches such as participatory technology assessment and integrated environmental assessment, indicate that the boundaries between science and policy have become obsolete. This development has been captured and justified by concepts such as transdisciplinarity (Gibbons et al. 1994) or postnormal science (Funtowicz and Ravetz 1993). In a sense, both concepts link the classic political ideal of learning through participation to the current notion that disciplinary academic inquiry is unable to cope with the huge complexities of social issues.

To what extent is participation in the production of knowledge able to do what is promised by participatory discourse? Where boundaries between science and policy are getting diffuse, the policy types discussed before constitute a context in which the rules of the game normally associated with policy tend to overrule one basic feature of knowledge production in science, the articulation and testing of rival hypotheses. From the angle of policy, especially the need for consensus on knowledge as to enable political consensus, there might be a discrepancy between promise and practice. However, in order to assess the possibilities for and limitations of participation as knowledge production, to point to political context only would not be convincing. After all, many scientists, policy-makers and policy stakeholders in society are genuinely interested in new forms of knowledge production. Quite some approaches, tools and procedures are in place to facilitate these efforts. Apart from the dominant context, these may make a difference.

Therefore, this section focuses on methods that are meant to facilitate participatory assessments. These participatory methods cover a range of approaches, tools and procedures developed in quite different traditions and fields. Without pretending to give an exact definition, this paper understands participatory methods as more or less precisely defined process steps and procedures for realizing a more or less precisely defined outcome, that can be distinguished from other (social science) methods in that groups of people are brought together at a specific location (which may also be at the www) in order to make some sort of assessment. To mention some: *Brain-*

200 KNOWLEDGE PRODUCTION AND THE LIMITS OF DEMOCRACY

storming, developed in the 1930s and 1940s (Osborn 1953) and *Focus Group methodology* have been developed in marketing. Methods known as *Simulation and Gaming* originate from the military and engineering (Parson 1996). *Policy delphi* (Linstone and Turoff 1975) and *backcasting* (Dreborg 1996). have their roots in future research and technology assessment; they were originally not meant to be participatory at all. The founding father of policy science, Lasswell (1960), developed the *Decision Seminar* in the early sixties. Social scientists, who participated in the democratic wave of the sixties and seventies, developed what has become known as deliberative methods, such as *Planungszelle* and *Citizens Forum* in Germany (Renn, forthcoming) and the *Dialectical Method* by Mason and Mitroff (1981) in the US. Controversies on issues related to science and technology in the 1970s and 80s gave rise to methods such as *scientific mediation* (Abrams and Primack 1980), Citizens' Juries (Seley 1983) and the *Consensus Conference* (Joss and Durant 1995).

It must be very clear that, even if all these methods maybe labelled participatory in some way, they are largely different in terms of their specific aim, scope and procedure. It is beyond the scope of this section to discuss the methods in detail or to give a judgment on their specific qualities.[3] However, using examples from specific procedures, I will show that participatory integrated environmental assessments, participatory technology assessments and similar exercises may suffer from mechanisms that prevent the articulation and assessment of conflicting knowledge claims.

What are the elementary requirements for an approach that aims at learning through participation? I would suggest that such an approach should have the following features: Provided that the relevant stakeholders have been adequately identified,

3. It must facilitate the interactive *articulation* of conflicting viewpoints, e.g., rival hypotheses and information;

4. It must facilitate the interactive *evaluation* of conflicting lines of argument, taking into account a wide range of aspects;

5. It must facilitate a *conclusion* of the debate, either in the form of consensus recommendations or of rival policy alternatives.

Any approach that meets these requirements might be fit for what I refer to as *problem structuring*. Drawing upon the work by Mitroff and Dunn, I define this concept as *the articulation, confrontation, comparison and, where possible, integration of as many contradictory arguments as possible*. It is assumed that an understanding of conflicting approaches is the key to policy learning (Figure 1, cell A). 'Policy as learning' relates to problem structuring and a reasoned problem choice (Hisschemöller 1993: 170; Hisschemöller and Hoppe 2001: 63). Problem structuring can be understood as second order learning (Fischer 1990: 248) or double-loop learning (Argyris and Schön 1978) in that all these concepts point to some sort of dialogue between actors who draw upon specific constructions of (social) reality.

From the perspective of mainstream participatory discourse, the idea of problem structuring through a dialogue between stakeholders with different perceptions of the issue might look common sense or even trivial. However, for critics of participation, the idea of a dialogue might look controversial or even dangerous. There seem to be three main objections against this idea: First, because the dialogue participants will not give up their core assumptions with respect to their key interests and basic values, a dialogue may very well yield an escalation of (latent) conflict. Second, as there is an almost infinite number of stakeholder views and selection of 'relevant views' can impossibly happen on objective grounds, a dialogue between all involved is not feasible. Third, since institutionalized voices, i.e., vested interests, have a huge advantage in terms of information and communication skills, a dialogue might lead to a situation in which the views already powerful get even more attention.

Methodological devices for stakeholder participation in the production of knowledge for policy might be understood as responses to these quite fundamental criticisms, as they try to avoid the risks of failure that the criticisms imply. My point is – and this will be illustrated below – that specific methodological devices may adequately address either one of these risks, but are unable to address them all at the same time. Moreover, the first and the third point may lead to inconsistent devices, as the obvious answer to a fear for escalation is to build in mechanisms for the avoidance or reduction of conflict, whereas the need to address the status quo inevitably leads to devices that, in a sense, encourage conflict.

Having made these general observations, I will now discuss the mechanisms (1) to reduce and (2) to enhance conflict. The second criticism, about the infinite number of claims, will be dealt with under 2), since it is especially relevant if one pretends to articulate the relevant viewpoints with respect to a given issue.

Mechanisms to Reduce Conflict

Learning through a dialogue between conflicting stakeholder views is not an easy exercise for those involved. This is because learning may touch upon rethinking and redefining ones interest, which is likely to yield a new perspective on reality with respect to the issue discussed (Connolly 1974). There are many situations indeed, where a dialogue may not work or even be counter-productive. And even in case one may not immediately expect an unwillingness on the side of stakeholders to listen to one another, the design of the process is critical. Many tools depart from the assumption that the main barrier for open and safe atmosphere is that participants get stuck within their daily routine. Tools are aimed at stimulating 'out of the box' thinking, which would enable participants to put themselves in the shoes of others. This would imply, however, that an immediate focus on conflicting views is avoided. There are basically three mechanisms that may help to provide trust.

The first mechanism is to reduce the heterogeneity of the stakeholder group to be involved. This may lead to a discussion among stakeholders who have a lot in common, such as culture, expertise, interest, place or age. Although a more or less homogeneous stakeholder group is not a guarantee for consensus and even if consensus is not explicitly aimed for, this mechanism increases the probability of a dialogue among like-minded stakeholders. It is fair to say that some homogeneity will always be needed to enable any dialogue at all, but it may reduce the learning potential when specific views are consciously excluded. The building of arguments in a like-minded group of citizens or stakeholder representatives maybe warranted as part of a process of broader interaction. It should however be noticed that participation is frequently meant to build some 'countervailing power' by groups of people that are considered to be in a position of disadvantage. In the 1970s and 1980s, participatory tools have been developed to specifically serve the purpose of developing informed citizens' considerations in addition to, or opposed to dominating expert views. The *consensus conference* can be considered an example of such an approach.

The second mechanism is to prevent any discussion. Dialogue participants are supposed to listen to one another and to react but not to criticize. The best examples are tools aimed at the identification of issues and options for problem solving. *Brainstorming* is meant to identify options for creative problem solving. People are supposed to mutually stimulate creative association. This requires a high tempo and, as Osborne clearly points out a well defined problem. The procedure does not allow for any discussion on the options raised. *Focus group methodology* is meant to identify issues of concern to people, but not for the structured exchange and exploration of conflicting views (Huitema et al. 2003). Many other tools, e.g,. backcasting (Van de Kerkhof 2004), although not explicitly aimed at preventing discussion, focus on identification rather than on exploring argument. A quite different example of a tool that may fit into this category is the original scope and focus of *policy Delphi.* This method explicitly aims at confronting and comparing conflicting lines of argument among experts. The experts do not talk with one another, though, they communicate through written statements via the facilitator.

The third mechanism is to allow for discussion, except on conflicting knowledge claims. The major example here is *Simulation and gaming*, which covers a wide range of tools and procedures. *Simulation and gaming* has originally been developed for assisting policy makers and risk managers to prevent group-think. This is done by taking a perspective distant from daily short-term routine, such as by putting the stakeholders in a position different from their own (e.g., of an opponent), a different place (China instead of Germany) or a different time (the future instead of the present). People are thus put in a role different from who they really are. Hence, they are prevented from putting forward their genuine concerns, their knowledge and their views.

However, even participatory methods that are explicitly aimed at enhancing a deliberative process use mechanisms to prevent conflicting knowledge claims from being considered. An example can be found in the participatory method proposal by Renn (forthcoming), which concentrates on conflicting values as a means to create a common discourse, thereby leaving the so-called 'cognitive aspects' to the experts.

Mechanisms to Enhance Conflict

It has been argued that social-constructivism predicts an infinite number of individual problem constructions. However, there is evidence to suggest that the range of problem constructions is actually quite limited and can be obtained by 15–30 interviews using repertory grid analysis (Dunn 2001, also for an interactive application).[4] What follows from this is that, for participatory assessments, a thorough stakeholder analysis might reveal the potentially conflicting views with respect to a certain issue.

As a quality indicator, good participatory assessments produce counter-intuitive results. Dunn further specifies this point where he argues:

> From the standpoint of communication theory and language, the information content of a hypothesis tends to be negatively related to its relative frequency, or probability of occurrence. Hypotheses that are mentioned more frequently – those on which there is substantial consensus – have less probative value than rarely mentioned hypotheses, because highly probable or predictable hypotheses do not challenge accepted knowledge claims (Dunn 2001: 425–6).

This observation suggests that the mechanisms should be in place to articulate assumptions that are marginal and build comparably strong cases for each line of argument. I would suggest that, in order to give marginal hypotheses a fair chance, there are three mechanisms that could be explored.

The first mechanism is, in contrast with Simulation and Gaming, to articulate and assess authentic conflict. The articulation and assessment of conflicting lines of argument has proven to be difficult and may depend on national custom. The Dutch experience reveals that a dialogue group, if not adequately facilitated, shows an inclination toward artificial consensus, i.e., agreement on an abstract level, leaving the 'hot issues' aside. In a similar vein, participatory assessments show many difficulties in selecting priorities. A tool such as the Devil's Advocate might help to articulate critical views. The weakness of such an approach appears to be that it replaces authentic conflict by artificial conflict, as participants maybe aware that the Devil's Advocate plays a role. Only authentic conflict provides persons debating a controversial issue with the stimulus to put forward genuine concerns and to articulate the knowledge and experience it draws upon. As comes forward social-psychological experiments, learning benefits from authentic conflict, but artificial conflict may reinforce stakeholders' original beliefs (Nemeth, Brown and Rogers 2001).

The second mechanism is to articulate and discuss stakeholders' taken for granted assumptions. As taken for granted assumptions are normally hidden below the surface of conscious reflection, the articulation of such assumptions may require a critical attitude and a lot of *Why* questions. Instead of shifting the discussion to values, Mason and Mitroff (1981) have suggested a *Dialectical Method*, which is based on the idea that a process of argument and learning not only requires that stakeholders get a better understanding of the views put forward by others, but also of their own. The *Dialectical Method* suggests that an articulation and assessment of conflicting claims and arguments supposes a shifting back and forth between heterogeneous and homogeneous groups. Also methods such as *Interactive Policy Delphi*, *Scientific Mediation* and *Citizens' Juries* may provide tools for articulating conflicting assumptions. One of the main problems is that, to my knowledge, detailed evaluations of these and other participatory methods in practice are scarce.

The third mechanism is to be transparent with respect to the quality of policy argument. Most of what has been written about this issue, is about the evaluation of public policies. Van de Kerkhof (2004), evaluating the Dutch stakeholder dialogue on Climate Options for the Long term (COOL) suggests to focus on *differentiation*, i.e., the range of different aspects that the stakeholders have taken into account, *empirical content*, i.e., as to whether they have used state of the art scientific knowledge and *integration*, i.e., the way different aspects and claims are linked in the conclusive arguments.

204   KNOWLEDGE PRODUCTION AND THE LIMITS OF DEMOCRACY

Conclusions on Methodology

In conclusion, methods used for participatory knowledge production address different requirements. In order to facilitate a dialogue between (potentially) conflicting views, the building of trust is necessary to get the process going, but mechanisms for building trust through 'out of the box thinking' are inconsistent with mechanisms to articulate and assess conflicting lines of argument. More work needs to be done, especially because mechanisms that exclude conflicting knowledge claims from consideration appear to be dominant and well documented evaluations of the methods and tools used in participatory assessments are scarce.

CONCLUSIONS: THE CHALLENGE FOR DEMOCRATIC INSTITUTIONS

The international environmental policy- and science communities face a growing awareness that problems in areas such as global environmental change require a participatory and transdisciplinary approach. National governments and the European Union are trying and experimenting with fora for inter- and transdisciplinary work. However, criticisms raised with respect to participatory knowledge production deserve more serious attention than they have so far received in the environmental policy sciences community. In many countries, participatory practices have not led to an increased public involvement in public policy. Instead, the gap between government and society even seems to widen. The explanation offered in this paper is that present day democracies lack institutions for managing conflicting knowledge claims, thereby defining *institutions* primarily as the formal and informal 'rules of the game' for reaching at political decisions.

The analysis of three policy types indicates that conflicting knowledge claims are organized out of the political process by transforming them into interests and values. I do not claim that there is anything wrong with policies that manage to work out solutions for conflicts of interests and values, as long as the stakeholders involved (either citizens or representatives from NGOs or private business) agree on what the values and interests are. What can be learned from environmental policy analysis though, is that in many cases knowledge input from scientific experts as well as non-scientists, is neglected. The dominant policy context tends to intrude into the domain of scientific knowledge production, either by turning knowledge claims into (vested) interest advocacy positions or by imposing scientific consensus in the interest of politics. Under these conditions, participation may become an obstacle for the advancement of policy-making, which may result in non-decisions with respect to urgent social and environmental issues. What remains, for the time being, is policy rhetoric, paper work and scientific discourse on Governance by Networking. What may come is an anti-participatory backlash, driven by the widely shared view that government is there to simply 'do the right thing,' but which is highly unlikely to effectively address urgent social issues either.

A look into the dominant policy types and their mechanisms to limit participation to interests and values, provides a picture which actually offers two main alternatives, i.e., the traditional strong monolithic government with low opportunity for public participation versus a multi-actor multi-level governance with a rather high level of (vested interest) participation. Remarkably, what is missing in the landscape of political institutions, is strong governance, which in my view also includes strong elected bodies, combined with a high level of public participation. This raises the question, as to whether such a model of democracy might be imaginable, and what conceptual barriers must be removed in order to present it as a visible and appealing alternative within the framework of existing policy-making institutions?

The discussion of participatory methods may provide some basic notions with respect to the direction of the institutional challenge. The first one is the concept of *problem structuring*. This concept is embedded in the idea that western culture, including western Europe and the United States, is *solution oriented* in that it focuses on developing (procedures for finding) solutions rather than specifying problems. Democracies have focused on providing methods for conflict resolution, e.g., through negotiation and accommodation, and have invested in applied science methodologies that might reduce decision costs. What has been neglected is the orientation toward *problem finding*, i.e., a focus on articulating and investigating into potentially rival positions instead of avoiding these. Rather than closure, such orientation would relate to the articulation and testing of rival hypotheses through involving knowledge from a variety of sources. The benefits of such an approach might largely outweigh the costs of symbolic policies and unimplemented decisions.

The second notion that might be critical in reflecting on institutions for addressing conflicting knowledge claims, draws upon the diverging positions with respect to the feasibility of stakeholder dialogue on conflicting lines of argument. On the one hand there is the position that the major barrier for such a dialogue is the difficulties people have with 'out of the box' thinking. From this perspective, it makes sense to shape the discussion in such a way that an immediate focus on authentic conflict is avoided. This may happen either by organizing more or less homogeneous groups ('consumers,' 'poor farmers,' etc.) or, in case of heterogeneous groups, by introducing mechanisms that create some distance between subject (the participant) and object (the issue for discussion). On the other hand, there is the position that the barriers for 'learning' not only originate from persons' lack of understanding of perspectives taken by others, but that it is especially difficult for persons to question their own taken for granted assumptions. The best way of doing this is with the 'help' of critics.

I tend to argue that both methodological positions are not irreconcilable and that the question how to structure a debate as to enable participants to engage in the process is an empirical one.

However, the notion that learning benefits most from authentic conflict seems to contradict common-sense. There is this widespread idea that persons are capable of a rational judgment with respect to the public good once they are brought into a *disinterested* position. This idea is found in 20th century political philosophy. A much cited example is the 'veil of ignorance,' introduced by Rawls (1971) in making a case for a political order that might be supported by rational persons irrespective of their specific position in society. Another well-known example is Habermas' (1981) notion of the 'ideal speech' situation, which would enable people to have an open conversation in the absence of power. Both concepts can be understood as methodological devices to address social dilemmas. As such they reflect a powerful notion found in democratic theories from Rousseau and J.S. Mill, that participatory democracies should resist partiality on the side of their citizens. Renn and Webler (1995) explicitly refer to the ideal-speech situation in developing devices for a fair and competent dialogue.

It is my observation indeed, that, certainly in the western European tradition, institutions for political participation and even some of the most well-known tools for participatory assessments, are (implicitly) based on the assumption that the success of joint problem solving is dependant on the readiness of persons involved to take a low profile with respect to their specific interests. From the social constructivist perspective, which is guiding my argument, I would suggest to turn this assumption upside down. A focus on diverging interests, once this happens openly, may facilitate a discussion on conflicting knowledge claims, because stakeholders do possess specific knowledge *because* of their interested position that other stakeholders for the same reason don't.

Hence, participatory policy analysis could assist in (re)shaping political institutions in such a way that they address the structuring of problems through encouraging the articulation of conflicting arguments and thereby take stakeholders as interested persons and groups who have, not in spite but because of their biased position, specific knowledge to offer.

Vrije Unversiteit, Amsterdam, The Netherlands

NOTES

[1] This section draws upon findings from earlier work, especially Hisschemöller 1993; Hisschemöller and Hoppe 1996/2001; Hisschemöller, Hoppe, Groenewegen and Midden 2001.

[2] By three Chambers consisting of scientists, artists, engineers and captains of industry.

[3] See for an overview and more detailed analysis Mayer 1997; Van de Kerkhof 2004.

[4] Elements, such as climate options, are combined into a number of triads. For each triad, the following questions are asked: (1) In what respect do two of these options equal one another and differ from the third? The answer to this question provides a construct, such as end of pipe versus innovative. (2) What would you prefer as a criterion for the long term? And (3) Please rank all options now on this dimension.

REFERENCES

Abrams, N.E. and J.R. Primack (1980), 'Helping the public decide, the case of radioactive waste management', *Environment* **22**, 3: 14–40.

Argyris, C. and D.A. Schön (1978), *Organisational Learning: A Theory of Action Perspective*, Reading, MA: Addison-Wesley.

Barker, A. and M. Couper (1984), 'The art of quasi-judicial administration. The planning appeal and inquiry systems in England', *Urban Law and Policy* **6**: 376–83.

Berelson, B.R., P.F. Lazarsfeld and W.N. McPhee (1954), *Voting*, Chicago: University of Chicago Press.

Braybrooke, D. and Ch.E. Lindblom (1963), *A Strategy of Decision*, New York: Free Press

Connolly, W.E. (1974), *The Terms of Political Discourse*, Lexington, MA: D.C. Heath and Company.

Dahl, R.A. (1985), *Nuclear Weapons: Democracy Versus Guardianship*, Syracuse, NY: SUNY Press.

Dreborg, K.H. (1996), 'The essence of backcasting', *Futures* **28**, 9: 813–28.

Dunn, W.N. (2001), 'Using the method of context validation to mitigate type III errors in environmental policy analysis', in M. Hisschemöller, R. Hoppe, W. s and J. Ravetz (eds.), *Knowledge, Power and Participation in Environmental Policy*, Policy Studies Review Annual Volume 12, New Brunswick, NJ: Transaction Publishers, pp. 417–36.

Ezrahi, Y. (1990), *The Descent of Icarus. Science and the Transformation of Modern Democracy*, Cambridge, MA: Harvard University Press.

Fischer, F. (1990), *Technocracy and the Politics of Expertise*, Newbury Park, CA: Sage Publications.

Funtowicz, S. and J. Ravetz (1993), 'Science for the post-normal age', *Futures* **25**, 7: 739–55.

Gibbons, M., C. Limoges, H. Nowotny, S. Schwartzman, P. Scott and M. Trow (1994), *The New Production of Knowledge*, London: Sage Publications.

Gupta, J. (2001), 'Effectiveness of air pollution treaties. The role of knowledsge, power and participation', in M. Hisschemöller, W.N. Dunn, R. Hoppe and J. Ravetz (eds.), *Knowledge, Power and Participation in Environmental Policy Analysis*, Policy Studies Review Annual Volume 12, New Brunswick, NJ: Transaction Publishers: pp. 145–75.

Habermas, J. (1981), *Theorie des kommunikativen Handelns*, Frankfurt a.M.: Suhrkamp.

Hajer, M.A. (1995), *The Politics of Environmental Discourse: Ecological Modernization and the Policy Process*, New York: Oxford University Press.

Hayek, F.A. (1944), *The Road to Serfdom*, Chicago, ILL: Chicago University Press.

Hisschemöller, M. (1993), *De democratie van problemen. De relatie tussen de inhoud van beleidsproblemen en methoden van politieke besluitvorming*, Amsterdam: VU Uitgeverij.

Hisschemöller, M., W.N. Dunn, R. Hoppe and J. Ravetz (2001), 'Knowledge, power and participation in environmental policy analysis: An introduction, in M. Hisschemöller, W.N. Dunn, R. Hoppe and J. Ravetz (eds.), *Policy Knowledge, Power and Participation in Environmental Policy Analysis*, Studies Review Annual Volume 12, New Brunswick, NJ: Transaction Publishers, pp. 1–28.

Hisschemöller, M. and R. Hoppe (2001), 'Coping with intractable controversies: The case for problem structuring in policy design and analysis', in M. Hisschemöller, W.N. Dunn, R. Hoppe and J. Ravetz (eds.), *Knowledge, Power and Participation Environmental Policy Analysis*, Policy Studies Review Annual Volume 12, New Brunswick, NJ: Transaction Publishers, pp. 47–72.

Hisschemöller, M., R. Hoppe, P. Groenewegen and C.J.H. Midden (2001), 'Knowledge use and political choice in Dutch environmental policy: A problem structuring perspective on real life experiments in extended peer review', in M. Hisschemöller, W.N. Dunn, R. Hoppe and J. Ravetz (eds.), *Knowledge, Power and Participation in Environmental Policy Analysis*, Policy Studies Review Annual Volume 12, New Brunswick, NJ: Transaction Publishers, pp. 437–70.

Hoppe, R. and A. Peterse (1993), *Handling Frozen Fire. Political Culture and Risk Management*, Boulder, CO: Westview Press.

Huitema, D. (2003), 'Siting unwanted land uses: Does interactive decision-making help?', in B. Denters, O. van Heffen, J. Huisman and P.J. Klok (eds.), *The Rise of Interactive Governance and Quasi-Markets*, Amsterdam: Kluwer Academic Publishers.

Huitema, D., M.F. van de Kerkhof and G.M. van Tilburg (2003), *River Dialogue. Focusing on the IJsselmeer. A Report of Nine Focus Groups on Water Management in the IJsselmeer Area for the River Dialogue Project*, IVM Report (W-03/22), Amsterdam: Vrije Universiteit.

Huntington, S.P. (1981), 'Reform and stability in a modernizing, multi-ethnic society', *Politikon* **8**: 8–26.

Jasanoff, S. (1990), *The Fifth Branch: Science Advisers as Policymakers*, Cambridge, MA: Harvard University Press.

Jasanoff, S. (2001), 'Civilization and madness: The great BSE scare of 1996', in M. Hisschemöller, W.N. Dunn, R. Hoppe and J. Ravetz (eds.), *Knowledge, Power and Participation in Environmental Policy Analysis*, Policy Studies Review Annual Volume 12, New Brunswick, NJ: Transaction Publishers, pp. 251–70.

Joss, S. and J. Durant (eds.), (1995), *Public Participation in Science: The Role of Consensus Conferences in Europe*, London: The Science Museum.

Lasswell, H.D. (1960), 'Technique of decision seminars', *Midwest Journal of Political Science* **4** (3): 213–33.

Lindblom, C.E. (1965), *The Intelligence of Democracy*, New York: Free Press.

Lijphart, A. (1968), *The Politics of Accomodation: Pluralism and Democracy in the Netherlands*, Berkeley, CA: University of California Press.

Lijphart, A. (1984), *Democracies. Patterns of Majoritarian and Consensus Government in Twenty-One Countries*, New Haven, CT: Yale University Press.

Linstone, H.A. and M. Turoff (eds.), (1975), *The Delphi Method, Technics and Applications*, Reading, MA: Addison-Wesley.

Mason, R.O. and I.I. Mitroff (1981), *Challenging Strategic Planning Assumptions. Theory, Cases, and Techniques*, New York: John Wiley & Sons.

Mayer, I. (1997), *Debating Technologies. A Methodological Contribution to the Design and Evaluation of Participatory Policy Analysis*, Tilburg, The Netherlands: Tilburg University Press.

Nemeth, C., K. Brown and J. Rogers (2001), 'Devil's advocate versus authentic dissent: Stimulating quantity and quality', *European Journal of Social Psychology* **31**: 707–20.

Osborn, A. (1953), *Applied Imagination: Principles and Procedures of Creative Problem Solving*, New York: Scribner.

Parson, E. (1996), *How Should We Study Global Environmental Problems? A Plea for Unconventional Methods of Assessment and Synthesis*, Laxenburg, Austria: IIASA Paper.

Rawls, J. (1971), *A Theory of Justice*, Cambridge, MA: Harvard University Press.

Renn, O. (forthcoming), 'The challenge of integrating deliberation and expertise. Participation and discourse in risk management', in T.L. McDaniels and M. Small (eds.), *Risk and Governance*.

Renn, O., T. Webler and P. Wiedemann (1995), *Fairness and Competence in Citizen Participation. Evaluating Models for Environmental Discourse,* Dordrecht, The Netherlands, Kluwer Academic Publishers.

Sabatier, P.A. and H. Jenkins-Smith (eds.), (1993), *Policy Change and Learning. An Advocacy Coalition Approach*, Boulder, CO: Westview Press.

Schön, D.A. (1983), *The Reflective Practitioner. How Professionals Think in Action*, New York: Basic Books.

Schumpeter, J.A. ([1942] 1976), *Capitalism, Socialism and Democracy*, New York: Harper & Row.

Schwarz, M. and M. Thompson (1990), *Divided We Stand: Redefining Politics, Technology, and Social Choice*, New York: Harvester & Wheatsheaf.

Seley, J.F. (1983), *The Politics of Public Facility Planning*, Lexington, MA: D.C. Heath and Company.

Van de Kerkhof, M.F. (2004), *Debating Climate Change. A Study on Stakeholder Participation in an Integrated Assessment of Long-Term Climate Change in the Netherlands*, Utrecht, The Netherlands: Lemma Publishing.

CHAPTER 12

SHEILA JASANOFF

JUDGMENT UNDER SIEGE: THE THREE-BODY PROBLEM OF EXPERT LEGITIMACY

The 2004 U.S. presidential election will be remembered for many things: the close margin of George W. Bush's victory in the electoral vote (he would have lost to the Democratic candidate, John Kerry, if only the state of Ohio had swung the other way); renewed questions about the viability of the electoral college; the inaccuracies of exit polling; and the stark division of the country's voting map into the "red" states of America's heartland and the "blue" states of its more cosmopolitan periphery. More curiously, it was also an election that pitted one perception of the relationship of science and government against another. On Kerry's side were multiple Nobel laureates and other leaders of the scientific community, vocally asserting that the Bush administration had betrayed science in the pursuit of crass political objectives.[1] These advocates cited the administration's lack of support for embryonic stem cell research, which many saw as the next great frontier in biomedicine; they also pointed to a series of White House actions manipulating or suppressing scientific data – on environment, public health, and defense – that the government had deemed inconsistent with its overall political strategy.[2] Against these charges, Republican representatives either issued denials or claimed a superior ethical sensibility, most explicitly so in George Bush' statement in the second presidential debate, "We've got to be very careful in balancing the ethics and the science...because science is important, but so is ethics, so is balancing life."[3]

This was not the way relations between science and government were scripted to work in mature democracies. For more than fifty years, cooperation, not friction, has been the order of the day in dealings between science and the state in technologically advanced nations. Indeed, the political scientist Etel Solingen predicted that there would be "happy convergence" between the goals of the state and its scientific communities, when there is "a high degree of consensus between state structures and scientists, who enjoy internal freedom of inquiry and relatively comfortable material rewards" (Solingen 1993: 43). More empirically minded researchers have shown that it is in the state's interest to sponsor scientists as a separate "estate" to assist in matters of policy formulation and implementation (Price 1965), a "brain bank" to draw on for policy legitimation (Boffey 1975), or a skilled and specialized labor force available to lend its authority to the state in times of national need (Mukerji 1989).

Sabine Maasen and Peter Weingart (eds.), Democratization of Expertise? Exploring Novel Forms of Scientific Advice in Political Decision-Making – Sociology of the Sciences, vol. 24, 209–224.
© Springer Science+Business Media B.V. 2009

These findings are consistent with the vision of a new social contract between science and the state put forward by presidential adviser Vannevar Bush at the end of the Second World War: in exchange for continued governmental support and freedom to define their research priorities and methods, scientists would provide the public with beneficial discoveries and a trained workforce (Bush 1945). Put succinctly, the contract provided money and liberty in exchange for knowledge and technical skills. In reality, the liberty offered to science was never complete; state support always came with strings attached, and the strings have both multiplied and tightened over the years, so that science today operates within a thick web of social constraints. Vannevar Bush's hope of weaning American science from dependence on military aims, and so liberating scientists from national security controls, for example, turned out to be illusory (Dennis 1994, 2004). Other state priorities, from environmental protection to enhanced university-industry collaboration, have shaped both the content and structure of governmental funding programs. And ethical concerns have led to varied restrictions on the use of federal funds for animal, human and biotechnological research, as well as a host of accounting and reporting mechanisms to force science to explain itself better to its public sponsors (Stokes 1997; see also Kevles 1998; Guston 2000).

Yet in a liberal democratic order, in which the state must continually expose itself to "attestive witnessing" by citizens (Ezrahi 1990), scientists' cooperation in national projects remains an invaluable resource, and states for the most part have been unwilling to risk serious breaks with organized science for the sake of short-term political gains. Rancorous partisan politics of the sort that surfaced in the 2004 presidential election is therefore unprecedented in the annals of recent science and seems contrary to the spirit of the postwar social contract. If scientists and their expertise are of such immense value, then mere party politics ought not to disrupt the peaceful coexistence of science and the state. Why, then, have relations between science and the party in power have soured of late? Why, more specifically, have tensions arisen around biomedical funding, for decades one of the most pampered and cosseted areas of U.S. science policy?

In addressing these questions, I argue that the implicit contract between science and the state has subtly shifted focus in recent decades. Although public support for science remains of paramount concern to researchers and research institutions, the politics of science no longer centers solely on the size of appropriations. Only by continually reaffirming its utility in expanding domains of application can science assert sustained claims on the public till. At stake, therefore, is a deeper right to define how, when, by whom, and to what extent science will be integrated into the solution of public problems, and who, indeed, will frame those problems in the first place. These questions straddle the line between science and politics, or truth and power, and attempts to answer them entail inevitable boundary conflicts over where the role of science ends and that of politics or policy begins (on boundary conflicts involving science see Gieryn 1999). Precisely this sort of boundary struggle can be discerned in George Bush's desire to locate the stem cell controversy in the domain of "ethics" and "balancing life" – areas of acknowledged political supremacy – rather than in "science."

As the stakes have shifted, so too has the content of the decisions for which the state relies on science. Across a wide range of contemporary policy issues, uncertainty and ignorance militate against the design of unambiguous technical solutions. Broadly characterized by the label of "risk" (Beck 1992), the threats that states are asked to mitigate on behalf of their citizens require the assessment of complex trajectories of social, technological and environmental change. There is typically no single, universally agreed upon, correct outcome to these sorts of assessments. Incoherence, not consensus, is the normal epistemological condition in many domains of policy-relevant knowledge.

In offering opinions on such contested and indeterminate issues, scientists can no longer stand on firmly secured platforms of knowledge. The questions contemporary policymakers ask of science are rarely of a kind that can be answered by scientists from within the parameters of their home disciplines. Scientists instead are expected to function as experts, that is, as persons possessing analytic skills grounded in practice and experience, rather than as truth-tellers with unmediated access to ascertainable facts. Accordingly, the technical expert's attributes often include, but are rarely limited to, mastery of a particular area of knowledge. What politicians and society increasingly expect from experts in decisionmaking processes is the ability to size up heterogeneous bodies of knowledge and to offer balanced opinions, based on less than perfect understanding, on issues that lie within nobody's precise disciplinary competence. Judgment in the face of uncertainty, and the capacity to exercise that judgment in the public interest, are the chief qualifications sought today from experts asked to inform policymaking. In these circumstances, the central question is no longer which scientific assessments are right, or even more technically defensible, but whose recommendations the public should accept as credible and authoritative. That question leads immediately to a second-order query: whose judgment should we trust, and on what basis?

All this has important consequences for democracy. So long as scientists were called upon mainly to provide specialized information – or, in the familiar phrase, to "speak truth to power" – there was no need to worry unduly about their political accountability. Peer pressure, it was assumed, would keep scientists honest; deviations from standards of professional rectitude would be uncovered and corrected by communities whose central function was to discover the truth and make it public. The shift from science to expertise, and from knowledge to judgment, confounds this easy expectation. Holding persons accountable for speaking the truth is different from holding them accountable for exercising judgment. And yet, as I show below, the discourses and practices of accountability have not yet caught up with the changing role of experts in the political process. Accountability measures in many societies still focus on one or possibly two of the three bodies that are relevant to the effective integration of science and politics: the bodies of knowledge that experts represent ("good science"); the bodies of the experts themselves ("unbiased experts"); and the bodies through which experts offer judgment in policy domains ("balanced committees"). The democratization of expertise demands, I suggest, renewed attention to the third of these bodies – namely, the institutions of advice-giving. It is this neglected level of analysis that I foreground in this paper, arguing that attempts to ensure data quality and lack of bias are not alone enough to serve the needs of democratic governance; measures are also needed for securing the legitimacy of expert advisory bodies.

To this end, I begin by briefly discussing the disjunction between the rhetoric of scientific disinterestedness in U.S. science policy and the reality of science's thickening ties to society. I then use two phases of the American debate on the peer review of regulatory science to show how a reductionist rhetoric of "good science" – encompassing only the first of the three relevant bodies – continues to dominate the U.S. framing of the problem of expert legitimacy. That framing, I show, is deeply resistant to counter-discourses emanating both from academic research in science and technology studies (STS) and from national regulatory practice. One consequence of that framing, in turn, is to blur the lines of expert accountability, drawing attention away from the institutional setting of advice-giving and concealing the need for public review of expert judgments.

Contrasting the American approach with that of Britain and Germany, I next illustrate how partial vision is not unique to the United States: these political cultures have also dealt selectively with the three-body problem, each highlighting one body at the expense of the others. I conclude by discussing the need for a richer theorization of the authority of policy-related expertise. Through that work we can begin to supplement, and compensate for, the weaknesses of accountability systems that reduce the three-body problem of expert legitimation to one or another of its constitutive elements.

THE DISINTERESTEDNESS OF SCIENCE: RHETORIC AND REALITY

It is tempting to dismiss the scientific community's opposition to the Bush administration in 2004 as the complaints of a disappointed suitor. As the veteran science journalist Daniel Greenberg has documented, scientists dependent on the state for research support now constitute a powerful lobby, no less insistent in their demand for public funds than the beneficiaries of any other entitlement program (Greenberg 2001). This dependence, according to Greenberg, has bred a variety of deplorable behaviors in the scientific community, ranging from overselling the promises of research to outright fraud. Scientists, on this account, have lost faith in an administration that has not simply poured funds into new research frontiers identified by their communities, from climate change to embryonic stem cells. Political success has eroded what Greenberg sees as science's historically pristine ethical position – a position famously characterized by the sociologist Robert Merton as including the virtues of openness, communal sharing of results, and lack of interest in the financial or political consequences of inquiry (Merton 1973).

The overt political positioning of prominent scientists and scientific organizations in the 2004 U.S. presidential campaign was certainly a stark reminder that the years of ivory-tower science, guided by the Mertonian norms, are definitively over. With active state encouragement,[4] scientists in the United States and around the world have become avid entrepreneurs, not only in the search for nature's secrets but also in tirelessly seeking support for their work before and after the phase of discovery. The resulting multi-level engagement of scientists with politicians, venture capitalists, journalists, the mass media, patent lawyers, the courts, and the public renders almost fantastic any residual notions of science's disinterestedness and detachment from society.

But the messiness of today's interactions between science and society is not news to academic observers of that relationship. At no point in the growth of modern science was detachment from society the norm (see, for instance, Shapin and Schaffer 1985; Golinski 1992; Jardine 1999; and for the modern period, Kevles 1987). Rather, science and other powerful social institutions – church, state, corporations, the media – have long engaged in negotiations about the nature and limits of the patronage that scientists enjoy, and the associated constraints on their liberty. Science's vaunted detachment, in other words, is a partial thing, achieved through societal interactions that are necessarily political. Galileo had to submit his beliefs formally to the strictures of the Catholic Church. Today, the controls on science are more subtle, if more pervasive: they relate, for the most part, not to scientists' substantive beliefs on particular issues, but to the means with which they are allowed to pursue certain lines of inquiry, the conditions under which their advice is sought, and the extent to which research trajectories are subordinated to political imperatives such as war or national security, environmental protection, or finding cures for life-threatening disease.

Clearly, then, it is both simplistic and ahistorical to claim that science became politicized for the first time at the turn of the 21st century, for arguably there never has been a time when the work of science was wholly distinct from the work of politics.[5] To be sure, substantial qualitative and quantitative changes have occurred in the performance of science and in its social, political, and economic links to society. Some have argued that the increased density of science-society interactions, particularly in the conduct of research, constitutes in and of itself a break with the past. European science policy scholars, in particular, have suggested that purely curiosity-driven, basic, or "Mode 1" research is a thing of the past. Instead, they say, we have entered the era of "Mode 2" science, characterized by wide-ranging interdisciplinarity, growing public-private collaboration, the rise of application-driven sciences, and increased demands for social accountability (Gibbons et al. 1994; Nowotny et al. 2001). These observations have rightly been seen as significant for the organization and funding of science, but their implications go further. Thoroughgoing changes in the production of science cannot but affect the foundations of scientific authority. As long as scientists could claim objective access to nature's laws, on the basis of observations unbiased by personal or political interests, that alone was sufficient to underwrite their expertise. With science more and more being produced in the service of social ends, the possibility of bias is far more evident, and the grounds of expert authority correspondingly in greater need of rearticulation.

Yet if the practices of science have evolved in the ways that scholars have documented, the political rhetoric around science has not kept pace, particularly in the United States. One looks in vain for explicit acknowledgment that expert deliberations are a site of hybrid judgment, combining technical and normative considerations. Instead, virtually all public pronouncements on the role of science in policy home in on the need for untainted science and the associated need to defend science from the corrupting encroachments of money and politics. Thus, the United States charged the European Union with maintaining an illegal and *unscientific* moratorium against the importation of genetically modified crops and foods in its 2004 case in the World Trade Organization (Winickoff et al., in press). In a related vein, Europe's commitment to the precautionary principle has been widely decried by U.S. critics as a politically motivated opt-out from the intellectual rigor of *scientific* risk assessment – not taken on board as a valid normative response to uncertainty. U.S. scientists for their part have also tended to frame disputes over policy-relevant science in the black and white language of purity and deviance, whose logic is to represent scientists as accountable only to their own specialist peers. The Union of Concerned Scientists, for example, focused its February 2004 pre-election campaign on the need to restore scientific integrity in policymaking.

This lag between reality and rhetoric does not advance the cause of democracy. If science has always been in some deep sense political, then it is not the *fact* of science's embeddedness in politics that should any longer be of primary concern, but rather the *nature* of that embedding and its implications for accountable governance. When an American administration withholds research funds from a promising area of biomedicine, or denies the validity of the scientific consensus on climate change, the problem is not the threat that is thereby posed to the mythic purity of science. Of greater importance is the tacit change that such disagreements signal in the rules of the game by which science and politics have previously ordered their relations vis-à-vis each other. There is an apparent retreat from politicians' earlier deference to scientists' judgments on basic elements of science policy: when is it in the public's best interests to fund a promising line of research; and when is contested knowledge robust enough to justify policy action? Put differently, what seems to have eroded in the Bush era is not so much the integrity of science itself as scientists' influence over decisions at the nexus of science and politics – above all, over how to deliberate and how to act when knowledge and understanding are incomplete. It is that shift in the seat of judgment that calls for analysis.

Occurring largely outside the purview of formal legal and political institutions, such struggles over the institutional division of power between science and politics raise important questions for governance and political theory. At a time when the vast majority of public decisions involve sizeable components of technical analysis, any change in the relative positions of scientific and political judgment carries with it a displacement in the exercise of power, with possible consequences for participation, deliberation and accountability. Now no less than in 1960s, when Yale University political theorist Robert Dahl used it as the title of his seminal treatment of democracy, the question at the heart of politics remains, "Who governs?" (Dahl 1961). A difference, however, is that technical decisionmaking is now more visibly and continuously a part of the playing field of politics. Consequently, there is a need to enlarge the scope of political analysis to take on board, or retheorize, the role of experts in processes of governance. A look at two episodes in some 25 years of debate on the quality of regulatory science in the United States underscores the need for conceptual advances.

THE RECURSIVE POLITICS OF REGULATORY PEER REVIEW

The quality and reliability of science for public policy have been recurrent themes in the United States for more than a quarter-century (see particularly Jasanoff 1990).

Critics of policy-relevant science have sought to ensure its robustness, and a favorite device has been the review of the government's findings and conclusions by other, appropriately trained eyes. This demand supplements the more general requirement of public justification, minimally through notice and comment provisions, that has been a part of the U.S. administrative process since the mid-1940s. On the assumption that policymakers' judgments on science as on other matters will be mission-oriented, and hence potentially biased, critics have demanded that those judgments be submitted to validation by experts, in other words, to peer review. Ongoing controversy over the forms of peer review in U.S. regulatory decisionmaking offers an ideal site for reconsidering the rules of accountability that secure expert legitimacy in that country. Two moments in the peer review debate are of particular interest, the first occurring in the 1980s and the second in 2003 and 2004. Together, they illustrate the power of a framing of policy-relevant science that persistently denies its hybridity and normative content.

An issue that captured the attention of U.S. policymakers perhaps more than any other in the late 1970s was what to do about cancer-causing substances in the environment (for a detailed account of these developments, see Brickman et al. 1985). In 1971, President Richard Nixon declared a "war on cancer," which resonated with public fears of an insidious and irreversible disease that had become, with heart disease, one of the country's two biggest killers. Federal agencies responsible for regulating the environment, pesticides, food and drugs, cosmetics, consumer products, and worker health and safety took up the challenge of working out principles for assessing and controlling the risks of carcinogens. Operating under newly precautionary legislation, these agencies were charged with preventing harms to public health and the environment before they materialized. In the case of carcinogens, this meant identifying the hazardous substances, if possible, before they entered the commercial pipeline or were dispersed into the environment. To carry out that preventive mandate, regulators felt they had to make many conservative assumptions: about the mechanisms of cancer causation (e.g., no safe threshold of exposure); dose-response relationships (e.g., that cancer incidence at high exposure doses should be linearly extrapolated to low doses); and the relationship between humans and test animals (e.g., that humans should be assumed to be similar to the most sensitive test animals). Affected industries argued, for their part, that these assumptions were scientifically untenable and led to irrational, economically burdensome regulation. Agency risk assessments, critics charged, would not hold up to scrutiny if they were peer reviewed by impartial experts with no ties to the agencies' regulatory mission.

It emerged in the ensuing debate that the term "peer review" was highly malleable and functioned effectively as an instrument of boundary maintenance between science and politics, as well as between regulators and their critics (Jasanoff 1987). Virtually all interested parties agreed that the science underlying regulatory decisions ought to be reviewed in some fashion, but there the consensus ended. There were disagreements about who the reviewers should be, what should be reviewed, and how review processes should be structured and organized. In my 1990 study of these developments, I concluded that "peer review," had fallen together with the more general function of expert advice-giving (Jasanoff 1990). Scientific advisory committees had become what I termed a "fifth branch" of government, and they functioned best when they conformed to standards of political legitimacy as well as technical rationality. Advisory processes produced the highest levels of participant satisfaction when they permitted the joint negotiation of technical and normative concerns and when expert advisers remained answerable to the publics affected by their judgments.

The peer review debate of the 1980s ended pragmatically in a victory for agency discretion and decentralized decisionmaking. An influential 1983 report by the National Research Council (NRC), the advisory arm of the National Academies, concluded, against industry advocacy to the contrary, that risk assessment functions should not be located within a single expert body but should rather be carried out separately by each relevant agency, consistent with its particular statutory mandate (National Research Council 1983). Called the Red Book because of its cover color, the report defined risk assessment as a purely technical activity, as distinct from risk management, a process taking account of economic and social factors. Yet background studies commissioned for the Red Book affirmed that risk assessment, too, was a hybrid process, calling for value judgments as well as technical analysis. Those findings buttressed the report's conclusion that risk assessment should remain within the control of authorized regulatory bodies – and, by extension, their legislative missions. Implicitly, the Red Book concluded that process and substance legitimately influence each other in regulatory analysis. While not cognizant of the academic literature in science and technology studies, the NRC report was in this respect compatible with emerging STS insights about the co-production of knowledge and norms (Jasanoff 2004).

In retrospect, we can say that the Red Book's practice was more sophisticated than its rhetoric, but – unreflexively adopted and with no theoretical underpinnings – the practice proved less influential than the rhetoric. Discursively, the report gave strong support to the characterization of risk assessment as a science, a view that powerfully informs regulatory discourse to this day. In terms of practice, the report offered a far more subtle view of the weaving together of analysis and judgment. In effect, the Red Book contained within its covers two contradictory views of risk

Table 1: Two discourses of risk analysis

Dominant Discourse	Insights from Regulatory Practice
Risk assessment (RA) should be separate from risk management (RM).	Judgment enters into both RA and RM; there can be no clear separation.
RA should not include economic, social, and political concerns.	RA occurs within particular frames which reflect social and political values and may differ across cultures.
RA can be and should be science-based.	RA is limited by uncertainty and ignorance.
There is a clear boundary between science and politics; there exist pre-established criteria by which we can decide whether an analysis is science-based.	The boundary between science and policy is not given in advance; criteria are established by negotiation and convention.

assessment and regulatory science that would come into clearer focus over subsequent years (see Table 1).[6] Politically, however, it was the less nuanced and more easily instrumentalized view that proved more durable.

As if to illustrate this point, a second major episode in the politics of U.S. peer review began unfolding in the summer of 2003. On August 29 of that year, the Office of Information and Regulatory Affairs (OIRA) of the Office of Management and Budget (OMB), the economic arm of the executive branch, issued a *Proposed Bulletin on Peer Review and Information Quality*. The *Bulletin*'s stated purpose was to ensure "meaningful peer review" of science pertaining to regulation, as part of an "ongoing effort to improve the quality, objectivity, utility, and integrity of information disseminated by the federal government."[7] Specifically targeted was the category of "significant regulatory information," that is, information that could have "a clear and substantial impact on important public policies or important private sector decisions with a possible impact of more than $100 million in any year." The proposal, it was estimated, would have far-reaching influence across the federal agencies, requiring 200 or more draft technical documents to be subjected annually to OMB-supervised "formal, independent, external" peer review (Anderson 2003).

The *Bulletin*'s principal intellectual justification was that the quality of science crucially depends on peer review. As the text observed,

> A "peer review," as used in this document for scientific and technical information relevant to regulatory policies, is a scientifically rigorous review and critique of a study's methods, results, and findings by others in the field with requisite training and expertise. Independent, objective peer review has long been regarded as a critical element in ensuring the reliability of scientific analyses. For decades, the American academic and scientific communities have withheld acknowledgment of scientific studies that have not been subject to rigorous independent peer review (*Bulletin*, Supplementary Information, 68 *Federal Register* 54024).

These statements, and indeed the entire thrust of the *Bulletin*, assumed that science is a unitary form of activity, that peer review likewise is a singular, well-defined process, and that the application of peer review to all forms of science – including regulatory science – can therefore be viewed as unproblematic. Peer review was advanced as a kind of objective audit mechanism for policy-relevant science, to be applied as a backstop to studies conducted by and for regulatory agencies. This characterization downplayed the political implications of removing ultimate control of the review process from the jurisdiction of the regulatory agencies to the OMB, and thereby to a White House with a notably anti-regulatory philosophy.

The *Bulletin* appeared to turn the clock back on years of policy learning. Not only was it oblivious to research findings on the interpretive flexibility of peer review, but it also went against the grain of the 1983 NRC Red Book in calling for a single, uniform process of validation, approved by OMB, for all types of regulatory science. The impulse toward standardization, overriding cross-agency differences in practice, was visible at many points in the proposal text, as exemplified by the following quotations:[8]

> 54024: "Existing agency peer review mechanisms have not always been sufficient to ensure the reliability of regulatory information disseminated or relied upon by federal agencies."

54024: "Even when agencies do conduct timely peer reviews, such reviews are sometimes undertaken by people who are not independent of the agencies."

54025: "When an agency does initiate a program to select outside peer reviewers for regulatory science, it sometimes selects the same reviewers for all or nearly all of its peer reviews on a particular topic."

54025: "it is also essential to grant the peer reviewers access to sufficient information…"

54025: "the results are not always available for public scrutiny or comment."

54025: "experience has shown that they are not always followed by all of the federal agencies, and that actual practice has not always lived up to the ideals underlying the various agencies' manuals."[9]

Not surprisingly, the OMB proposal came under severe criticism from many quarters, including the highest reaches of organized science, where the move to draw regulatory peer review within the supervisory ambit of an already suspect executive branch was immediately perceived as political. In November 2003, the National Academy of Sciences hosted a public workshop at which were aired many research and practice-based objections to the proposal. By mid-December, the end of the official comment period on the proposed *Bulletin*, 187 written responses had been filed, some two-thirds critical of the proposal. At its February 2004 annual meeting, the American Association for the Advancement of Science (AAAS) adopted a resolution calling on OMB to withdraw the proposal. Reasons offered by AAAS and other opponents included fears of political interference, unnecessary bureaucratic hurdles, asymmetric treatment of experts funded by agencies and corporations (the proposal initially identified only the former as having a potential conflict of interest), and the rigidity of a "one size fits all" approach to review (see, for example, Steinbrook 2004; *Philadelphia Inquirer,* January 25, 2004).

For me personally these developments posed particular intellectual challenges. As an STS scholar whose work had specifically addressed the topic of regulatory peer review, I had a stake in opposing a policy initiative that seemed inconsistent with the basic findings of my and my colleagues' work. I was also aware that my own study of advisory committees could be, and had been, uncritically read as an endorsement of more stringent peer review, with little attention to my observations about the constructedness of policy-relevant knowledge.[10] Breaking a lifetime habit of standing apart from current controversies, I therefore participated in the National Academy workshop and, more exceptionally, submitted written comments to OMB urging that the proposal be retracted. My conclusions that regulatory science is different in context and content from research science, and that "peer review" therefore cannot be uncritically translated from one domain to the other, were referenced in the AAAS resolution and to some extent reported in the media. Their impact on OMB, however, proved slight.

On April 15, 2004, OMB issued a substantially revised proposal, taking note of many of the submitted comments.[11] The new version narrowed the scope of the most stringent peer review requirement to a newly defined category of "influential scientific information" containing, as a subset, "highly influential scientific assessments"; it also granted more flexibility to agencies to design their peer review procedures, and it removed the one-sided restriction on experts whose research was funded by regulatory agencies. At the core, however, the proposal continued to embrace the notion of an autonomous science whose quality and objectivity could be improved in a straightforward way through critical scrutiny by "peers." Instructively, the revised proposal cited my work on advisory committees only to support the propositions that peer review practices are varied and that fair and rigorous review can build consensus around agency actions based on science. That regulatory science is, by its very nature, a site of politics was evidently inconsistent with the deeply entrenched Mertonian discourse of science's integrity, independence, quality and rigor. In this case, as we have seen, the discourse of scientific integrity masked a profoundly political institutional realignment between regulators and the White House. Neither scholarship nor practical wisdom was able to undermine a discourse that offered such substantial instrumental benefits to the ruling interests of the moment.

CULTURAL PRACTICES OF EXPERT LEGITIMATION

As in the United States, regulators in Britain and Germany have accepted risk assessment as a principled approach to ordering knowledge and weighing policy alternatives, and risk analysis occupies a central place in both countries' practices for coping with the consequences of technological change.[12] Yet in neither European national setting has the methodological robustness of risk assessment received nearly the same attention as in the United States, and nowhere else have political battle lines been drawn around the design of regulatory peer review. Tacitly, at least, decision-making in both European countries takes on board the hybrid picture of risk judgments that represented one face of the 1983 NRC Red Book report (see Table 1). That hybridity, in turn, demands accountability to wider interests than those of relevant technical communities – forcing consideration of more than simply the body of policy-related knowledge. Accordingly, political representation remains part and parcel of the process of risk analysis in both countries, consciously built into the design of expert committees and consultative processes.

But even though the hybridity of risk judgments is generally conceded, practices for ensuring lack of bias remain partial and untheorized, reflecting different cultural traditions for the construction of public knowledge – traditions that I have elsewhere termed "civic epistemology" (Jasanoff 2005: chapter 10). On the whole, the focus in British regulatory circles is on the body of the expert: accountable judgment is sought through consultation with persons whose capacity to exercise judgment on the public's behalf is regarded as superior, even privileged. Though members of British expert panels can and do represent both technical specialties and social interests, ultimately it is the excellence of each person's individual discernment that the state most crucially relies on. To a remarkable extent the legitimacy of British expertise remains tied to the person of the individual expert, who achieves standing not only through knowledge and competence, but through a demonstrated record of service to society. It is as if the expert's function is as much to discern the public's needs and to define the public good as to provide appropriate technical knowledge and information for resolving the matter at hand.

Needless to say, this faith in individuals' power to see for the people could hardly exist in a more diverse or less empiricist cultural context, where common norms of judging and assessing facts were felt to be lacking. A cost of the British stress on virtuous expert bodies has been to protect the assumption of common vision itself from critical examination. Consequently, a narrow group of experts can with the best will in the world make erroneous judgments on matters that were too complex for their collective reckoning. Britain's infamous "mad cow" disaster of the 1990s illustrated the hazards of blind faith in embodied expertise at the expense of due consideration to what experts know, or can know, and the institutional context in which they exercise their expertise.[13]

In Germany, by contrast, expert committees are usually constituted as microcosms of the potentially interested segment of society; judgments produced in such settings are seen as unbiased not only by virtue of the participants' individual qualifications, but even more so by the incorporation of all relevant viewpoints into a collective output. Reliance on personal credentials is rare in Germany unless it is also backed by powerful institutional supports. To be an acknowledged expert in Germany, one ideally has to stand for a field of experience larger than one's own particular domain of technical mastery. And it is ultimately the institutional context for forming communal expert judgments that matters most to producing social robustness.

The constitution of such bodies reflects something important about what counts as right reason in the German public sphere. The painstakingly representative character of German expert advisory bodies, their membership often specified in detail by legislation, encodes a belief that it is possible to map the terrain of reason completely; an accurately configured map can then be translated into an institutionalized instrument of decisionmaking. An expert within such an institution functions almost an ambassador for a recognized region or place from among the allowable enclaves of reason. Rationality, the ultimate foundation of political legitimacy in Germany, flows from the collective reasoning produced by authoritatively constituted expert bodies. A paradoxical consequence of this map-making approach to public reasoning is that expert bodies, once constituted, leave no further room for *ad hoc* citizen intervention. They become perfectly enclosed systems, places for a rational micro-politics of pure reason, with no further need for external accountability to a wider, potentially excluded, and potentially irrational, public.

These contrasts help throw the cultural specificity of U.S. legitimation practices, and their solution to the three-body problem, into sharper relief. Professional skills and standing count for more in the United States than the intangible qualities of individual judgment (as in Britain) or institutional representation and balance (as in Germany). In a meritocracy that prides itself on individualism and objective markers of intelligence (Carson 2004), the surest way to become an expert is by climbing the ladder of professional recognition. What an expert stands for or has achieved outside the spheres of method and knowledge is of lesser consequence. Civic virtue is not a prime desideratum in the appointment of experts, although the capacity for team work obviously plays a part in the nomination and selection of experts for important advisory positions.

Of course, U.S. policy is not wholly insensitive to possible imbalances in the constitution of expert groups. The Federal Advisory Committee Act seeks to correct for just this eventuality through its requirement that committees be balanced in terms of the views they represent. Nonetheless, the dominant discourse of policy-relevant science remains unwaveringly committed to Mertonian ideals of purity and detachment, despite all scholarly demonstrations of hybridity and co-production. It is the perceived deviation from the transcendent objectivity of science that most often threatens expert legitimacy in the United States. Allegations that experts have been captured by political interests or by politically motivated research programs erupt in U.S. policy debates with a regularity unheard of in other modern democracies.

None of the three ideal-typical solutions to the problem of expert legitimacy provides for systematic lines of accountability running from experts to wider publics. Intensely political choices of individual experts and groupings remain concealed behind divergent national rhetorics and practices of accountability.

THEORY AS INTERVENTION: REGROUNDING THE LEGITIMACY OF EXPERTISE

Experts have become indispensable to the politics of nations, and indeed to transnational and global politics. Experts manage the ignorance and uncertainty that are endemic conditions of contemporary life and pose major challenges to the managerial pretensions and political legitimacy of democratically accountable governments. Faced with ever-changing arrays of issues and questions – based on shifting facts, untested technologies, incomplete understandings of social behavior, and unforeseen environmental externalities – governments need the backing of experts to assure citizens that they are acting responsibly, in good faith, and with adequate knowledge and foresight. The weight of political legitimation therefore rests increasingly on the shoulders of experts, and yet they occupy at best a shadowy place in the evolving discourse of democratic theory.

I have suggested that expert legitimacy should be reconceptualized as a three-body problem that pays explicit attention to each of the three bodies involved in producing expert judgments: the body of knowledge that experts concededly bring to decisionmaking; the individual bodies of the experts themselves; and the institutionalized bodies through which they offer judgment and policy advice. A brief study of the peer review debate in the United States illustrates the political hazards of too great an emphasis on the first body: the knowledge component of expert judgments. Coupled to an outmoded and uncritically accepted discourse of scientific purity, that emphasis has impeded wide debate by American scholars and publics on the credibility of experts and the institutional foundations of their legitimacy.

A brief contrast with two European political systems shows that the U.S. approach, while possibly unique in its commitment to a transcendental notion of scientific integrity, is not unique in the partiality of its understanding of expert legitimacy. The U.K. emphasis on the embodied expert and the German preoccupation with rational expert collectives each militates against deeper questioning of the constituents of expert authority. More specifically, no national decisionmaking system has as yet taken on board the fundamental STS insight that experts *construct* – they do not simply *find* – the knowledge base on which they rest their hybrid analytic-deliberative judgments. In each democratic society, then, an imperfect framing of the problem of expertise has foreclosed the continuous dialogue between expert and critical lay judgment that is imperative under contemporary conditions of ignorance and uncertainty.

Addressing this deficit in democratic practice requires us to recast the role of experts in terms that better lend themselves to political critique. Key to this move, as I have argued elsewhere, is to import notions of delegation and representation into the analysis of expert decisionmaking (Jasanoff 2003). Under a theory of delegation, experts can be seen as acting not only in furtherance of technical rationality, but also on behalf of their public constituencies, under cognitive and normative assumptions that are continually open to wider review. Equally, citizens need to recognize that governmental experts are there to make judgments on behalf of the common good rather than as spokespersons for the impersonal and unquestionable authority of science. In turn, this means that a full-fledged political accountability – looking not only inward to specialist peers but also outward to engaged publics – must become integral to the practices of expert deliberation.

We come, finally, to a concluding word on the role of scholarship and the relations of theory to practice. The history of expertise as a public problem in the United States and elsewhere suggests that deep reform – aimed not just at current policy practice but at its entrenched ideological foundations – cannot be effectively mounted at the surfaces of already framed debates and controversies. The long U.S. conversation on regulatory peer review illustrates the impediments to making critical voices heard within the press of politics as usual. To challenge, let alone change, deepseated habits of mind and thought, embedded in resistant institutional practices, requires the would-be critic of expert rule to step out and away from the four corners of ongoing disputes. It calls for the tacit assumptions of the workaday political world to be made explicit, and for new languages to be elaborated to describe previously unseen or taken-for-granted realities. Scholarship provides the platform for such intervention, and the power of the word, backed by historical knowledge and critical analysis, stands ready to be embraced in the project of rejuvenating democracy.

Harvard University, Cambridge, MA, USA

NOTES

[1] For a summary of these charges, see the statement on "Restoring Scientific Integrity in Policymaking" issued by the Union of Concerned Scientists on February 18, 2004, http://www.ucsusa.org/ (visited January 2005). See also US House of Representatives, Committee on Government Reform (Minority Report), *Politics and Science in the Bush Administration*, http://www.house.gov/reform/min/politicsandscience/pdfs/pdf_politics_and_science_rep.pdf (visited April 2004).

[2] The Republican strategy included placating the religious right on issues relating to abortion (hence, by extension, stem cell research), as well as industrial special interests opposed to stringent controls on carbon emissions and other forms of environmental regulation.

[3] CBS News.com, Text of Bush-Kerry Debate II, St, Louis, Missouri, October 8, 2004, http://www.cbsnews.com/stories/2004/10/08/politics/main648311.shtml (visited November 2004).

JUDGMENT UNDER SIEGE

[4] A notable example of such encouragement in the United States was the 1980 Bayh-Dole Act, which in effect required publicly funded researchers to seek commercial returns from their work. For critical accounts of the consequences of that legislation, see Press and Washburn (2000); Krimsky (2003).

[5] For more on the deep linkages between the construction of scientific and political power, see particularly Jasanoff (2004).

[6] Not all of the insights in the right-hand column, to be sure, were apparent to the authors of the Red Book. In particular, issues of framing and cross-cultural variation in risk assessment surfaced in these terms only in subsequent scholarly research, some of which used the Red Book and its assumptions as primary data for analysis. See, for example, Jasanoff 1986; Krimsky and Golding 1992).

[7] *Proposed Bulletin on Peer Review and Information Quality* (hereafter cited as *Bulletin*), Summary, 68 *Federal Register* 54023, September 15, 2003.

[8] All page citations are to the *Federal Register*, vol. 68, no. 178 (September 15, 2003).

[9] I am indebted to John Mathew and John Price for identifying these extracts.

[10] It was not the first time my work had been misread in the policy domain as affirming rather than critiquing dominant conceptions of the science-policy relationship. Other similar episodes included a misinterpretation of my work on science advice in a U.S. Supreme Court decision on the admissibility of expert evidence. See Jasanoff (1996).

[11] http://www.whitehouse.gov/omb/inforeg/peer_review041404.pdf (visited January 2005).

[12] The regulation of biotechnology provides an especially instructive site for observing national practices of regulatory practice and expert legitimation in action. See Jasanoff 2005.

[13] In April 2000, the U.K. government estimated that the total cost of the BSE crisis to the public sector would be 3.7 billion pounds by the end of the 2001-2002 fiscal year. *The Inquiry into BSE and variant CJD in the United kingdom* [hereafter cited as *The Phillips Inquiry*] (2000), Volume 10, Economic Impact and International Trade, http://www.bseinquiry.gov.uk/report/volume10/chapter1.htm#258548 (visited April 2004).

REFERENCES

Anderson, F.R. (2003), 'Peer review of data', *The National Law Journal*, September 29, 2003.

Beck, U. (1992), *Risk Society: Towards a New Modernity*, London: Sage.

Boffey, P.M. (1975), *The Brain Bank of America: An Inquiry into the Politics of Science*, New York: McGraw Hill.

Brickman, R., S. Jasanoff, and T. Ilgen (1985), *Controlling Chemicals: The Politics of Regulation in Europe and the U.S.*, Ithaca, NY: Cornell University Press.

Bush, V. (1945), *Science – The Endless Frontier*, Washington, DC: US Government Printing Office.

Carson, J. (2004), 'The merit of science and the science of merit', in S. Jasanoff (ed.), *States of Knowledge: The Co-Production of Science and Social Order*, London: Routledge, pp. 181–205.

Dahl, R.A. (1961), *Who Governs?*, New Haven, CT: Yale University Press.

Dennis, M.A. (1994), '"Our first line of defense"': Two university laboratories in the postwar American State', *Isis* 85(3): 427–55.

Dennis, M.A. (2004), 'Reconstructing sociotechnical order: Vannevar Bush and US Science Policy', in S. Jasanoff (ed.), *States of Knowledge: The Co-Production of Science and Social Order*, London: Routledge.

Ezrahi, Y. (1990), *The Descent of Icarus: Science and the Transformation of Contemporary Democracy*, Cambridge, MA: Harvard University Press.

Gibbons, M., C. Limoges, H. Nowotny, S. Schwartzman, P. Scott, and M. Trow (1994), *The New Production of Knowledge*, London: Sage Publications.

Gieryn, T. (1999), *Cultural Boundaries of Science: Credibility on the Line*, Chicago: University of Chicago Press.

Golinski, J. (1992), *Science as Public Culture: Chemistry and Enlightenment in Britain, 1760–1820*, Cambridge, MA: Cambridge University Press.

Greenberg, D.S. (2001), *Science, Money, and Politics: Political Triumph and Ethical Erosion*, Chicago: University of Chicago Press.

Guston, D.H. (2000), *Between Politics and Science: Assuring the Integrity and Productivity of Research*, New York: Cambridge University Press.

Jardine, L. (1999), *Ingenious Pursuits: Building the Scientific Revolution*, London: Little, Brown.

Jasanoff, S. (1986), *Risk Management and Political Culture*, New York: Russell Sage Foundation.

Jasanoff, S. (1987), 'Contested boundaries in policy-relevant science', *Social Studies of Science* **17**: 195–230.

Jasanoff, S. (1990), *The Fifth Branch: Science Advisers as Policymakers*, Cambridge, MA: Harvard University Press.

Jasanoff, S. (1996), 'Beyond epistemology: Relativism and engagement in the politics of science,' *Social Studies of Science* **26**(2): 393–418.

Jasanoff, S. (2003), '(No) Accounting for expertise?', *Science and Public Policy* **30**(3): 157–62.

Jasanoff, S. (ed.) (2004), *States of Knowledge: The Co-Production of Science and Social Order*, London: Routledge.

Jasanoff, S. (2005), *Designs on Nature: Science and Democracy in Europe and the United States*, Princeton, NJ: Princeton University Press.

Kevles, D. (1987), *The Physicists: The History of a Scientific Community in Modern America*, Cambridge, MA: Harvard University Press.

Kevles, D.J. (1998), *The Baltimore Case: A Trial of Politics, Science, and Character*, New York: W.W. Norton.

Krimsky, S. (2003), *Science in the Private Interest: How the Lure of Profits Has Corrupted the Virtue of Biomedical Research*, Lanham, MD: Rowman-Littlefield, 2003).

Krimsky, S. and D. Golding (eds.), (1992), *Social Theories of Risk*, London: Praeger.

Merton, R.K. (1973), 'The normative structure of science,' in R.K. Merton, *The Sociology of Science: Theoretical and Empirical Investigations*, Chicago: University of Chicago Press, pp. 267–78.

Mukerji, C. (1989), *A Fragile Power: Scientists and the State*, Princeton, NJ: Princeton University Press.

National Research Council (1983), *Risk Assessment in the Federal Government: Managing the Process*, Washington, DC: National Academy Press.

Nowotny, H., P. Scott, and M. Gibbons (2001), *Re-Thinking Science: Knowledge and the Public in an Age of Uncertainty*, Cambridge, MA: Polity.

Philadelphia Inquirer (January 25, 2004), Editorial, 'The White House vs. Science.'

Office of Management and Budget (2003), *Proposed Bulletin on Peer Review and Information Quality*, Federal Register, Vol. 68, No. 178, Monday, September 15, pp. 54023–29.

Press, E. and J. Washburn (2000), 'The kept university', *Atlantic Monthly*, March 2000: 39–54.

Price, D.K. (1965), *The Scientific Estate*, Cambridge, MA: Harvard University Press.

Shapin, S. and S. Schaffer (1985), *Leviathan and the Air-Pump: Hobbes, Boyle, and the Experimental Life*, Princeton, NJ: Princeton University Press.

Solingen, E. (1993),'Between markets and the state: Scientists in comparative perspective,' *Comparative Politics* **26**: 31–51.

Steinbrook, R. (2004), 'Peer review and federal regulations', *New England Journal of Medicine* **350**(2): 103–4.

Stokes, D.E. (1997), *Pasteur's Quadrant: Basic Science and Technological Innovation*, Washington, DC: Brookings Institution.

Winickoff, D., S. Jasanoff, L. Busch, R. Grove-White, and B. Wynne (2005), 'Adjudicating the GM food wars: Science, risk, and democracy in world trade law', *Yale Journal of International Law* **30**: 81–123.

LIST OF AUTHORS

Alexander Bogner, Dr., Institute of Technology Assessment of the Austrian Academy of Sciences, Strohgasse 45, A-1030 Vienna, Austria.

Mark B. Brown, Assistant Professor, Department of Government, California State University, Sacramento, 6000 J Street, Sacramento, CA 95819-6089, USA.

Heather Douglas, Assistant Professor, Department of Philosophy, University of Tennessee, 808 McClung Tower, Knoxville, Tennessee, 37996-0480. USA.

David H. Guston, Professor of Political Science, Consortium for Science, Policy, and Outcomes, Arizona State University, PO Box 874401, Tempe, AZ 8528704401, USA.

Willem Halffman, Dr., School of Business, Public Administration and Technology, University of Twente, PO Box 217, 7500 AE Enschede, The Netherlands.

Harald Heinrichs, Professor Dr., Juniorprofessor for 'Sustainable Development and Participation,' Institute for Environmental and Sustainability Communication, D-21335 Lüneburg, Germany.

Matthijs Hisschemöller, Dr., Institute for Environmental Studies (IVM), Vrije Universiteit, De Boelelaan 1087, 1081 HV Amsterdam, The Netherlands.

Robert Hoppe, Professor Dr., School of Business, Public Administration and Technology, University of Twente, PO Box 217, 7500 AE Enschede, The Netherlands.

Sheila Jasanoff, Pforzheimer Professor of Science and Technology Studies, John F. Kennedy School of Government, Harvard University, 79 John F. Kennedy Street, Cambridge, MA, USA.

Simon Joss, Dr., Director of the Centre for the Study of Democracy (CSD), University of Westminster, London, UK.

Justus Lentsch, Dr., Institute for Science & Technology Studies (IWT) Bielefeld University, PO Box 100131, D-33501 Bielefeld, Germany.

Sabine Maasen, Professor Dr., Wissenschaftsforschung/Wissenschaftssoziologie, Science Studies, Universität Basel, Missionsstr. 21, CH-4003 Basel, Switzerland.

Wolfgang Menz, Dipl. Soz., Institut für Sozialforschung, Senckenberganlage 26, D-60325 Frankfurt am Main, Germany.

Frank Nullmeier, Professor Dr., Centre for Social Policy Research, Parkallee 39, D-28209 Bremen, Germany.

Stephen Turner, Professor, Department of Philosophy, University of South Florida, 4202 East Fowler Avenue, FAO 226, 33620 Tampa, Florida, USA.

Peter Weingart, Professor Dr., Institute for Science and Technology Studies (IWT), Bielefeld University, PO Box 10 01 31, D-33501 Bielefeld, Germany.

BIOGRAPHICAL NOTES

Alexander Bogner, born 1969, researcher at the Institute of Technology Assessment of the Austrian Academy of Sciences and lecturer at the University of Vienna. He studied sociology at the Universities of Salzburg and Frankfurt am Main. Post-graduate studies and researcher at the Institute for Advanced Studies in Vienna. His research focuses on science and technology studies, biopolitics and methods of empirical social research.

Mark B. Brown is an Assistant Professor in the Department of Government at California State University, Sacramento. He received a Ph.D. in political science from Rutgers University in 2001, and held a two-year postdoctoral fellowship at the Institute for Science & Technology Studies at Bielefeld University. His research interests are in the areas of democratic theory, modern and contemporary political theory, and science and technology studies. His current work focuses on the relationship between scientific and political representation.

Heather Douglas is currently Assistant Professor in the Department of Philosophy at the University of Tennessee after having spent several years as Philip M. Phibbs Assistant Professor of Science and Ethics at the University of Puget Sound. Her research focuses on the relationships among science, values, and policy-making, and has been supported by the National Science Foundation. Her work has been published in *Philosophy of Science, American Philosophical Quarterly, Synthese* and *Foundations of Chemistry*. She received her Ph.D. from the History and Philosophy of Science Program at the University of Pittsburgh in 1998.

David H. Guston is Professor of Political Science at Arizona State University and Associate Director of ASU's Consortium for Science, Policy, and Outcomes. His book *Between Politics and Science: Assuring the Integrity and Productivity of Research* (Cambridge U. Press, 2000), received the 2002 Don K. Price Prize by the American Political Science Association for best book in science and technology policy. He is co-editor of the forthcoming *Science, Technology, and Public Policy: The Next Generation of Research* (with D. Sarewitz, University of Wisconsin Press), co-author of *Informed Legislatures* (with M. Jones and L. M. Branscomb, University Press of America 1996), and co-editor of *The Fragile Contract* (with Ken Keniston, MIT Press 1994). Professor Guston is North American editor of the peer-reviewed journal *Science and Public Policy*. He holds a B.A. from Yale and a PhD from MIT, and he performed post-doctoral training at Harvard's Kennedy School of Government.

Willem Halffman studied social and political sciences in Antwerp, sociology at the Free University of Brussels and at Columbia University, New York, with specialisation in sociology of science. For ten years he worked at the Department of Science and Technology Dynamics at the University of Amsterdam, where he wrote a dissertation (March 2003) about the development of ecotoxicological knowledge for the regulation of environmental hazards of chemical substances in the Netherlands, England and the US. Meanwhile, he taught various courses related to science and technology studies and the relations between experts and policy makers. He cooperated on various advisory projects on environmental hazards and the use of insights of science and technology studies in policy-making. Since 2001 he has been working at the section Policy Sciences of the faculty of Business, Public Administration, and Technology of the University of Twente, in *Rethinking*, an extensive comparative research project on the relation between experts and decision makers, funded by the Dutch Science Foundation.

Harald Heinrichs, born 1970, is Juniorprofessor for *Sustainable Development and Participation* at the Institute for Environmental and Sustainability Communication, University of Lüneburg, Germany. Before, he spent several years as PhD student and Postdoc in the Program Group *Humans, Technology, Environment* at the Research Center Jülich, as visiting scholar at Tufts University in the US, and as lecturer at the University of Düsseldorf. His research and teaching focuses on processes of communication and interpretation regarding sustainable development, integration of (scientific) expertise into social contexts, and participatory procedures.

Matthijs Hisschemöller is Associate Professor at the Institute of Environmental Studies of the Vrije Universiteit in Amsterdam, where he leads the programme Tools for Transition. His main research relates to the study of participation in (inter)national environmental policy with a focus on the transition to a sustainable energy system, and the development and application of methods for participatory integrated environmental assessment. Among his publications is the Policy Studies Review Annual (Volume 12, 2001) *Knowledge, Power and Participation in Environmental Policy Analysis*, co-edited with Rob Hoppe, William Dunn and Jerome Ravetz.

Robert Hoppe (1950) holds an M.A. in Political Science (1974) from Catholic University Nijmegen. In 1974-1986 he worked at the Free University Amsterdam. From this university he earnt a Ph.D. in the social sciences in 1983. In 1986-1997 Hoppe was affiliated to the University of Amsterdam's Faculty of Political and Socio-Cultural Sciences, as professor and chair of Public Administration. In 1992-3 he spent a year as visiting professor and Fulbright Fellow at Rutgers University. Since 1997 he is professor and chair of Policy Studies at the Faculty of Business, Public Administration and Technology, University of Twente. He is co-editor and co-author (with Matthijs Hisschemöller, Bill Dunn and Jerry Ravetz) of *Knowledge, Power, and Participation in Environmental Policy Analysis* (Policy Studies Review Annual, Vol. 12, 2001). His present research interests are *boundary work arrangements* (he is program director of *Rethinking*, an extensive comparative research program on the relation between experts and decision makers, funded by the Dutch National Science Foundation), *policy-oriented learning* and the *argumentative turn* (in policy analysis), and *cultural theory*.

Sheila Jasanoff is Pforzheimer Professor of Science and Technology Studies at Harvard University's John F. Kennedy School of Government. She has held faculty appointments or visiting positions at several universities, including Cornell, Yale, Oxford, and Kyoto. At Cornell, she founded and chaired the Department of Science and Technology Studies. She has been a Fellow at the Berlin Institute for Advanced Study (*Wissenschaftskolleg*) and Resident Scholar at the Rockefeller

BIOGRAPHICAL NOTES 229

Foundation's study center in Bellagio. Her research centers on the role of science and technology in the authority structures of modern democratic societies, with a particular focus on the use of science in legal and political decision making. Her books on these topics include *The Fifth Branch* (1990) and *Science at the Bar* (1995). Jasanoff has served on the Board of Directors of the American Association for the Advancement of Science and as President of the Society for Social Studies of Science.

Simon Joss is Director of the Centre for the Study of Democracy (CSD), University of Westminster, London. He gained his PhD in political science from Imperial College, University of London, for his research on science and technology public policy. At CSD since 1997, he has led various research programmes focusing on socio-political aspects of science, technology and the environment, including the recent multinational *Public Accountability of Technology* project. His recent publications include contributions in: *Genetically Engineered Plants: Decision-Making under Uncertainty* (I. Taylor, ed., Haworth Press, NY, forthcoming); *Wozu Experten? Form und Funktion wissenschaftlicher Politikberatung in gesellschaftstheoretischer und empirischer Perspektive* (A. Bogner & H. Torgersen, eds, Westdeutscher Verlag, 2004); *The Bulletin of Science, Technology & Society* (2002); *Parliaments and Technology* (N. Vig & H. Paschen, eds., SUNY, 2000); and *Science and Public Policy* (1999).

Justus Lentsch is researcher and scientific coordinator of the interdisciplinary working group on *Scientific Advice to Policy in Democracy* at the Berlin-Brandenburg Academy of Sciences (BBAW) and at the Institute for Science & Technology Studies (IWT), Bielefeld University. Before, he held a two-year postdoctoral fellowship at the IWT. Having studied mathematics, physics and philosophy he received a philosophical doctorate (PhD) with a thesis on C.S. Peirce. Moreover, he is involved in a critical edition of Peirce's Lowell Lectures in cooperation with the American Peirce Edition Project. His current work focuses on the study of scientific expertise and advice to policy.

Sabine Maasen is Professor of Science Studies at the university of Basel. Her training is in sociology, psychology, and linguistics and her research interests are dynamics of knowledge, including processes of scientification. She has published several articles and books in the sociology of science and knowledge including *Biology as Society, Society as Biology. Metaphors* (with E. Mendelsohn and P. Weingart, Kluwer 1994), *Metaphors and the Dynamics of Knowledge* (Routledge 2000, with P. Weingart), and *Die Genealogie der Unmoral. Therapeutisierung sexueller Selbste* (*Genealogy of the Immoral. Therapeutic Constructions of Sexual Selves*, Suhrkamp 1998). Recently, she studies the social construction of volition and consciousness (e.g., *Voluntary Action. On Brains, Minds, and Sociality*, Oxford 2002, with W. Prinz & G. Roth). Moreover, she focuses on instantiations of governmentality in the conduct of selves-in-society, in new forms of knowledge production and in participatory science-policy arrangements.

Wolfgang Menz, born 1971, research assistant at the Institut für Sozialforschung at the Johann Wolfgang Goethe-Universität Frankfurt am Main and scholarship holder of the Hans-Böckler-Stiftung. He studied sociology at the universities of

Marburg, Edinburgh and Frankfurt. His research interests include the sociology of work and organization, science studies and methods of qualitative research.

Frank Nullmeier is a Professor of Political Science and Director of the Centre for Social Policy Research at the University of Bremen. He obtained his PhD from the Department of Political Science at the University of Hamburg. Since 2003, he is a member of the German Research Foundation's (DFG) Collaborative Research Center 597 *Transformations of the State*. His work focuses on social policy analysis and the role of (scientific) knowledge in decision making processes. He has also published books on welfare state legitimation and public administration. Some recent publications include: 2003: (with Tanja Pritzlaff und Achim Wiesner) *Mikro-Policy-Analyse. Ethnographische Politikforschung am Beispiel Hochschulpolitik*, Frankfurt a.M. and New York: Campus; 2003: (ed. with Matthias Leonhard Maier/Achim Hurrelmann/Tanja Pritzlaff/Achim Wiesner): *Politik als Lernprozess? Wissenszentrierte Ansätze in der Politikanalyse*, Opladen: Leske + Budrich.

Stephen Turner, born 1951, is Graduate Research Professor in the Department of Philosophy at the University of South Florida, Tampa, where he is also affiliated with the Department of Management. His Ph.D. is from the University of Missouri. He is the author of a number of books in the history of social science, philosophy of social science, and social and political theory, and organizational studies, and has also written extensively in science studies, especially on patronage and the politics and economics of science. His most recent book is *Liberal Democracy 3.0: Civil Society in an Age of Experts*.

Peter Weingart studied economics and sociology at the universities of Freiburg, Berlin and Princeton, holds a chair for sociology of science at the University of Bielefeld. He was director of the Center for Interdisciplinary Research (ZiF) 1989–1994, Fellow of the Wissenschaftskolleg 1983/84, Visiting Scholar at Harvard University 1984/85, Research Scholar at the Getty Research Institute 2000, since 1996 Visiting Professor at the University of Stellenbosch (South Africa), member of the Berlin-Brandenburg Academy of Sciences, and heads the Institute for Science and Technology Studies at Bielefeld (since 1994). He has published numerous books and articles in the sociology of science and science policy, i.a. *Die Stunde der Wahrheit?* (Velbrück, 2001), and *Von der Hypothese zur Katastrophe. Der anthropogene Klimawandel im Diskurs zwischen Wissenschaft, Politik und Massenmedien* (with A. Engels and P. Pansegrau, Opladen: Leske + Budrich, 2002).

AUTHOR INDEX

Aarts, W. 149
Abbott, A. 138, 149
Abels, G. 2, 10, 17
Abrams, N.E. 200, 206
AEBC Agriculture and Environment Biotechnology Commission 173, 175–6, 178–9, 182, 184, 186–7
Allum, N. 186
Altenhof, R. 86–89, 98
Andersen, I.-E. 94, 98
Anderson, F.R. 217, 223
Argyris, C. 200, 206
Austin, J.L. 123, 134
Avellan, S. 17

Badura, B. 44, 58
Bakker, W. 136, 149
Bal, R. 135, 149
Banthien, H. 172, 186
Barker, A. 41, 58, 136, 150, 194, 206
Basset, P. 149
Bauer, M.W. 186
Bechmann, A. 46, 58
Bechmann, G. 41, 48, 58
Beck, U. 22–3, 39, 143, 149, 211, 223
Becker, A. 39
Beemer, F.A. 144, 149
Beierle, T.C. 47, 60
Beker, M. 14–1, 149
Bell, D. 42, 58
Bellucci, S. 17, 83, 92, 99, 172, 187
Berelson, B.R. 194, 207
Beyme, K. 90, 98
Bijker, W. 149
Bimber, B.A. 2, 6, 17, 97–8
Bingham, E. 60
Boden, L. 60
Boehmer-Christiansen, S. 45, 58
Boffey, P.M. 209, 223
Bogner, A. 21, 38–9
Bohmann, J. 42, 58
Boiko, P. 158, 168
Bonß, W. 22–3, 39
Bora, A. 10, 17
Both, G. 149

Brais, N. 17
Braß, H. 89, 98
Braun, D. 77–8
Braybrooke, D. 194, 207
Brickman, R. 44–45, 55, 58, 136, 149, 215, 223
Brint, S. 16–17
Bröckling, U. 18
Brooks, H. 63, 78
Brown, K. 203, 208
Brown, M.B. 51, 59, 81, 98
Brown, P. 182, 186
Bruder, W. 44, 59
Bulletin, Supplementary Information 217, 223
Bundeskanzleramt 26, 39
Burgess, J. 176, 186
Busch, L. 224
Busenberg, G. 159, 168
Bush, V. 209–10, 212, 214, 222–3

Cabbage, M. 118–120
CAIB – *Columbia Accident Investigation Board* 103–5, 116–20
Calorie Control Council 71–2, 78
Caplan, N. 44, 59
Carson, J. 220, 223
Carson, L. 92, 99
Cassel, S. 41, 59
Caswill, C. 77–8
Centraal Bureau voor de Statistiek 142, 148–9
Centraal Planbureau 138–9, 149
Clapp, R. 60
Cohen, J. 11, 17
Colburn, T. 58, 59
Cole, S.A. 144, 149
Collins, H.M. 148–9
Columbia Accident Investigation Board CAIB/NASA 103, 112, 120
Connolly, W.E. 97, 99, 201, 207
Council of Science and Technology Advisors – CSTA 45, 59
Couper, M. 194, 206
Cozzens, S.E. 58–9

Cross, A. 12, 17
Crozier, M. 24, 39
Cruikshank, B. 15, 17
Crumpacker, W. 59
Cummings, L.C. 67, 78
Cunningham, P. 59
Curry, D. 182, 186

Dahl, R.A. 64, 78, 171–2, 183, 185–6, 193, 207, 214, 223
Daily Mail 174, 177, 186
De Gier, E. 149
De Marchi, B. 12, 18
De Wit, B. 141, 151
Den Boer, M.C. 144, 149
Deng, F.M. 12, 18
Dennis, M.A. 210, 223
Dennis, R. 59
Dente, B. 14, 17
Department for Environment Food and Rural Affairs – DEFRA 175–6, 179, 182, 186
Department of Health and Human Services – DHHS 68, 78
Derrida, J. 126, 134
Després, C. 5, 17
Deutscher Bundestag 89, 91, 99
Dewey, J. 168
Dienel, P.C. 92, 99
Dijstelbloem, H. 140, 145, 148–9
Douglas, H. 153, 155–6, 168
Dreborg, K.H. 200, 207
Dumanoski, D. 59
Dunn, W.N. 189, 200, 202–3, 207
Durant, J. 26, 39, 43, 59, 93, 99, 165, 168, 186, 200, 207
Durkheim, E. 42, 59

Edwards, A. 15, 17
Ellul, J. 1, 17
ENDS 174, 186
Engels, A. 19
Environmental Protection Agency – EPA/*Scientific Advisory Board* – SAB 45–6, 48, 50, 52, 59
EU-Commission 2, 12, 17
European Communities 172, 186
Expertisecentrum LNV 141, 149
Ezrahi, Y. 82, 99, 189, 191, 193, 207, 210, 223

Fareri, P. 17
Faustman, E. 168
Feenberg, A. 40, 82, 99
Felsch, A. 25, 39
Felt, U. 58, 59
Fineberg, H. 157, 169
Fiorino, D. 154, 168
Fischer, F. 154, 164, 168, 193, 200, 207
Fisher, F. 171, 186
Fitch, S. 59
Fixdahl, J. 173, 186
Flanagan, K. 59, 149
Flynn, J. 168
Forest Row GM Debate 180–1, 183, 186
Foucault, M. 15, 17
Frewer, L.J. 11, 18, 92, 100, 154, 169
Friedberg, E. 24, 38–9
Funtowicz, S. 77, 79, 199, 207
Futrell, R. 160–1, 167–8

Garnett, M.D. 6, 18
Gaskell, G. 173, 186
Gbikpi, B. 172, 186
Gellner, W. 41, 59
Gerber, M. 161, 168
Geuskens, I. 149
Gibbons, M. 2, 17–18, 39, 43, 59–60, 136, 149, 199, 207, 213, 223–4
Gieryn, T.F. 58–9, 135, 149, 210, 223
Gilpin, R. 1, 17
Gloede, F. 93, 99
Glynn, S. 41, 59, 147, 149
Gmeiner, R. 38–9
Golding, D. 223–4
Goldman, L.R. 60
Golinski, J. 213, 223
Gottweis, H. 26, 39
Greenberg, D.S. 212, 224
Groenewegen, P. 206–7
Grosch, D.J. 120
Grote, J.R. 172, 186
Grove-White, R. 224
Grundahl, J. 93, 99
Grunwald, A. 48, 58, 90, 98–9
Gupta, J. 197, 207
Guston, D.H. 2, 8, 17, 63, 74, 77–9, 94, 98–9, 165, 168, 210, 224
Habermas, J. 1, 17, 24, 39, 44, 47, 59, 82–3, 99, 123–4, 128–9, 134, 206–7
Hails, R. 175, 185–6
Hajer, M.A. 143, 149, 191, 207

AUTHOR INDEX

Halfacre, A.C. 14, 17
Halffman, W. 44, 59, 135–6, 150
Halliwell, J.E. 45, 59, 61
Halvorsen, T. 107, 120
Hamlett, P.W. 11, 12, 17
Hammond, K.R. 45, 59
Hampel, F. 89, 99
Harwood, W. 105–7, 120
Hayek, F.A. 191, 207
Heinrichs, H. 41–2, 58–9, 147, 150
Hemerijck, A.C. 140, 150
Hendriks, R. 149
Hennen, L. 83, 93, 99
Heuberger, P.S.C 149
Hilgartner, S. 58, 59
Hisschemöller, M. 189, 197, 200, 206–7
Hitzler, R. 82, 99
Hoffmann-Riem, W. 87, 99
Holzinger, K. 123, 134
Hoppe, R. 135–6, 139, 145, 150, 189,
    197, 200, 206–7
Hoppin, P. 60
Hörning, G. 93, 99
House of Commons 174–6, 185–6
Hronszky, I. 41, 58
Huber, P. 71, 79
Huijs, S. 145, 150
Huitema, D. 194, 202, 207
Huntington, S.P. 196, 207

Ibarreta, D. 149
Ilgen, T. 58, 149, 223
Independent on Sunday 174, 187
Inglehardt, R. 43, 59
In't Veld, R. 145, 150
Irwin, A. 162–3, 168
Ismayr, W. 87–9, 99

Jaeger, B. 94, 98
Jäger, W. 43, 59
James, D. 108, 112, 120
Janssen, J. 144, 150
Janssen, P.H.M. 149
Japp, K.P. 23, 39
Jardine, L. 213, 224
Jasanoff, S. 8, 13–14, 17, 44, 58–9, 65,
    74, 77, 79, 83–4, 99, 135, 149–50,
    193, 195, 197, 207, 209, 214–16, 219,
    222–4
Jaspers, M. 186
Jenkins-Smith, H. 195, 208

Joss, S. 26, 39, 43, 59, 83, 92–4, 99, 165,
    168, 171–2, 185–7, 200, 207

Kaiser, J. 71, 79
Kaiser, M. 14, 18
Kaplan, L. 161, 169
Kauffman, G.B. 67, 79
Keenan, M. 149
*Kenniscentrum Stedelijke Vernieuwing –
    KEI* 144, 150
Kenniscentrum Grote Steden 144, 150
Kenniston, K. 2, 17
Ketting, E. 144, 150
Kettner, M. 15, 17
Kevles, D. 210, 213, 224
Kinerlerer, J. 175, 186
Kinney, A. 161–2, 169
Kleimann, B. 43, 59
Klink, E. 138, 150
Klüwer, L. 11, 17
Knorr-Cetina, K. 16–17, 58, 60
Köbben, A. 142, 150
Kooiman, J. 172, 187
Körtner, U. 38–9
Krämer, S. 123, 134
Krasmann, S. 18
Kreibich, R. 43, 60
Krevert, P. 42, 46, 60, 86, 100
Krimsky, S. 43, 58, 60, 223–4
Kronje, G. 140, 150
Krücken, G. 10, 17
Kuhlmann, A. 21, 39
Kuhn, T.S. 112, 120, 154, 169
Küpper, W. 25, 39

Lacey, H. 154, 169
Lahsen, M. 12, 17
Laird, F. 154, 169
Lakoff, S.A. 1, 17
Langendonck, R. 149
Langewiesche, W. 106, 116, 119–20
Lapp, R.E. 1, 17
Lasswell, H.D. 200, 207
Latour, B. 37, 39
Lazarsfeld, P.F. 207
Lee, L. 155, 169
Lemke, T. 16, 18
Lentsch, J. 59
Leschine, T. 161–2, 169
Lester, J.P. 46, 60
Leuginger, J. 186

Leusner, J. 118–20
Libertore, A. 77, 79
Ligteringen, J. 17
Lijphart, A. 193, 196, 208
Limoges, C. 17, 59, 207, 223
Lindblom, C.E. 194, 207–8
Lindsey, N. 186
Linstone, H.A. 200, 208
Loeber, A. 149
Loka Institute 92, 100
Long, R.C. 47, 60
Longino, H. 168–9
Luhmann, N. 23, 38–9, 41–2, 60

Maasen, S. 1, 14, 18, 58, 60
MacRae, D.J. 135, 150
Majone, G.D. 13, 18
Mannheim, K. 43, 60
Mansbridge, J. 84, 98, 100
Marcus, A.I. 67, 79
Marcuse, H. 1, 18
Markle, G. 99
Martin, B. 58, 60, 92, 99
Mason, R.O. 200, 203, 208
Matheny, A. 17
Mayer, I. 206, 208
Mayntz, R. 45, 60, 131, 134
McCubbins, M.D. 78–9
McGinley, L. 71, 79
McPhee, W.N. 207
Menz, W. 21, 38–9
Merton, R.K. 212, 224
Meyer, R. 89, 97, 100
Meyers, J.P. 59
Meynaud, J. 1, 18
Michaels, D. 50, 60
Midden, C.J.H. 206–7
Mill, J.S. 101, 113–14, 120
Miller, P. 15, 18
Ministerie van Landbouw, N.e.V. 141, 150
Ministerie van Sociale Zaken en Werkgelegenheid 141, 150
Ministerie van Verkeer en Waterstaat 141, 150
Mitroff, I.I. 200, 203, 208
Mohr, A. 185, 187
Monforton, C. 60
Moore, J.A. 174, 187
Morrill, R. 168
Morris, N. 77, 79

Mukerji, C. 209, 224
Mumpower, J.L. 59
Murswieck, A. 6, 18, 41, 45, 60

Nationaal Dubo Centrum 144, 150
National Research Council 157, 169, 216, 224
National Toxicology Program – NTP 63, 66, 68, 79
Naumann, J. 92, 100
Neidhardt, F.M. 15, 18
Nelkin, D. 58, 60
Nemeth, C. 203, 208
Nielsen, T. H. 40
Nowotny, H. 2, 6, 13, 15, 17–18, 37, 39, 43–4, 59–60, 63, 74, 79, 207, 213, 223–4
Nullmeier, F. 25, 39, 123

O'Connor, J. 156, 169
O'Malley, P. 4, 16, 18
Office of Management and Budget – OMB 217, 224
Office of Technology Assessment – OTA 79
Oh, C.H. 44–5, 60
Oldersma, G.J. 137, 150
Omenn, G. 168
Ooijens, M. 149
Orlando Sentinel 103, 106, 110, 120
Ortmann, G. 25, 39
Osborn, A. 200, 208
Ott, K. 14, 18
Overleg Directeuren Planbureaus 139, 150
Oxford Economic Research Associates Ltd. – OXERA 45, 60
Ozawa, C. 159, 169
Ozonoff, D. 60

Paardekooper, C. 140, 150
Pansegrau, P. 19
Parau, C. 185, 187
Parson, E. 200, 208
Parsons, T. 42, 60
Paschen, H. 98–100
Pennington, H. 115, 120
Perrow, C. 117, 120
Petermann, T. 89, 98, 100
Peters, B.G. 41, 58, 136, 150, 172, 187
Peters, J. 90, 98, 100

AUTHOR INDEX

Peterse, A. 197, 207
Petersen, J.C. 83, 99–100
Petts, J. 186
Philadelphia Inquirer 218, 224
Pianin, E. 119–20
Pielke, R. Jr. 77, 79
Pierre, J. 172, 187
Pinch, T.J. 100, 148–49
Pitkin, H.F. 82, 84, 100
Plotke, D.. 82, 100
Press, E. 223–4
Price, D.K. 1, 9, 18, 209, 223–4
Priebe, P.M. 67, 79
Primack, J.R. 200, 206
Pritzlaff, T. 39

*Raad voor Ruimtelijk Milieu- en Natuur
    Onderzoek* 145, 150
Radaelli, C. 11, 18
Ramdharie, S. 142, 150
Ravetz, J. 136, 150, 189, 199, 207
Rawls, J. 205, 208
Reinecke, W.H. 12, 18
Renn, O. 43–6, 60, 98, 100, 135–6, 150,
    154, 163–4, 168–9, 200, 202, 206, 208
Renner, A. 186
Rich, R.F. 44–5, 60
Richards, E. 58, 60
*Rijksinstituut voor Volksgezondheid en
    Milieu* 139, 145, 150
*Rijksinstituut voor Volksgezondheid en
    Milieu/Milieu- en NatuurPlanbureau*
    145, 151
Rip, A. 52, 60
Robbins, A. 60
Rogers, J. 203, 208
Rooney, P. 168–9
Rose, N. 15–16, 18
Rosenbaum, W. 14, 17
Rosenstock, L. 155, 169
Rowe, G. 11, 18, 92, 100, 154, 169
Rayner 186

Sabatier, P.A. 195, 208
Sabel, C. 11
Saretzki, U. 8, 18, 92, 100
Sarewitz, D. 77, 79
Sawyer, K. 107, 120
Schaffer, S. 213, 224
Schelsky, H. 1, 18, 22, 24, 39
Schicktanz, S. 92, 100

Schimank, U. 43, 60
Schmitt, C. 102, 120
Schmitter, P. 172, 187
Schön, D.A. 191, 200, 206, 208
Schouw, G. 144, 150
Schreuder, A. 142, 151
Schulz, H.J. 39
Schumpeter, J.A. 191, 194, 208
Schüssel, W. 25–6, 39
Schuyt, C.J.M. 140, 145, 149
Schwartz, T. 78–9
Schwartzman, S. 17, 59, 207, 223
Schwarz, M. 196, 208
Sclove, R.E. 15, 18, 162, 165, 169
Scott, P. 17–18, 39, 59–60, 207, 223–4
Searle, J.R. 123, 125, 134
Sebaldt, M. 43, 60
Seley, J.F. 200, 208
Shapin, S. 135, 151, 213, 224
Shaw, G. 118–120
Skorupinski, B. 14, 18
Smith, B.L.R. 84, 92, 94, 100
Smith, G. 84, 92, 94, 100
Smith, W. 45, 59, 61
Sociaal Wetenschappelijke Raad 142,
    151
Solingen, E. 209, 224
Spear, K. 120
St. Petersburg Times 103, 120
Stehr, N. 42–3, 49, 60–1
Steinbrook, R. 218, 224
Sterling, A. 186
Stern, P.C. 157, 169
Stokes, D.E. 210, 224
Stolberg, S.G. 72, 79
Stone, D.A. 6, 18
Sutter, B. 16, 18

Tacke, V. 23, 39
Taschwer, K. 59
Taylor, I. 171, 187
Thompson, M. 196, 208
Thorpe, C.R. 102, 119–20
Thunert, M. 86, 100
Tiemann, H. 132, 134
Torgersen, H. 26, 39
Touraine, A. 1, 18
Trommelen, J. 142, 150
Tromp, H. 142, 150
Trow, M. 17, 59, 207, 223
Turner, S. 64, 77, 79

Turner, S.P. 101–2, 113, 120–1
Turoff, M. 200, 208
Tweede Kamer 146, 151
U.S. House of Representatives 77, 79
U.S. Senate 67, 69, 79

Van Asselt, M.B.A. 149
Von Asten, F. 149
Van Belle, G. 168
Van de Kerkhof, M.F 202–3, 206–8
Van de Peppel, R. 14, 18
Van den Berg, H. 138, 149
Van den Bogaard, A. 138, 149
Van der Giessen A. 149
Van der Meulen, B. 77, 79
Van Eijndhoven, J. 149
Van Est, R. 144, 149
Van Tilburg, G.M. 207
Vaughan, D. 117–18, 121
Van Waarden, F. 136, 149
Vogel, D. 136, 151

Wachelder, J. 162–3, 169
Wagenaar, H. 143, 149
Wakeford, T. 186
Wales, C. 93–4, 100
Walker, P.D. 120
Walmsley, M. 59
Warren, M. 98, 100
Washburn, J. 223–4

Weber, M. 16, 19, 22, 24, 39, 42, 61,
  102, 121
Webler, T. 60, 100, 169, 206, 208
Weill, C. 12, 19
Weinberg, A. 52, 58, 61
Weingart, P. 1, 9–10, 19, 21, 37, 40–1,
  44, 59, 61, 81, 83, 97, 100
Weiss, C.H. 44, 61
Weldon, S. 173, 187
Welz, W. 43, 59
Wetenschappelijke Raad voor het
  Regeringsbeleid 137, 139, 145, 151
Whittington, D. 135, 150
Wiedemann, P. 60, 100, 169, 208
Wiesner, A. 39
Willke, H. 38, 40
Windeler, A. 39
Wingens, M. 44, 61
Winickoff, D. 214, 224
Winner, L. 26, 40
Wirth, U. 126, 134
Woodhouse, E.J. 58–9
Wright, R. 1, 17
Wynne, B. 43, 61, 173, 186–7, 224

Young, I.M. 84, 100

Zhou, Z. 10, 19
Zilleßen, H. 43,

LaVergne, TN USA
04 September 2009
156951LV00002B/16/P